Fire Officer's Guide to

MANAGEMENT *and* LEADERSHIP

A Scenario-Based Approach

Jeffrey R. Barlow
Fire Chief
Burlington Fire Protection District
Burlington, Kentucky

JONES & BARTLETT
LEARNING

World Headquarters
Jones & Bartlett Learning
5 Wall Street
Burlington, MA 01803
978-443-5000
info@jblearning.com
www.jblearning.com

Jones & Bartlett Learning books and products are available through most bookstores and online booksellers. To contact Jones & Bartlett Learning directly, call 800-832-0034, fax 978-443-8000, or visit our website, www.jblearning.com.

Substantial discounts on bulk quantities of Jones & Bartlett Learning publications are available to corporations, professional associations, and other qualified organizations. For details and specific discount information, contact the special sales department at Jones & Bartlett Learning via the above contact information or send an email to specialsales@jblearning.com.

Production Credits
General Manager: Kimberly Brophy
VP, Product Development: Christine Emerton
Executive Editor: William Larkin
Editorial Assistant: Jessica Sturtevant
Director of Marketing: Brian Rooney
VP, Sales, Public Safety Group: Matthew Maniscalco
Associate Production Editor: Robert Furrier
VP, Manufacturing and Inventory Control: Therese Connell
Composition: S4Carlisle Publishing Services
Cover and Text Design: Kristin E. Parker
Rights & Media Specialist: Thais Miller
Cover Image (Title Page, Part Opener, Chapter Opener): Courtesy of Jeff Barlow
Printing and Binding: Edwards Brothers Malloy
Cover Printing: Edwards Brothers Malloy

Library of Congress Cataloging-in-Publication Data
Names: Barlow, Jeffrey, 1970- author.
Title: Fire officer's guide to management and leadership : a scenario-based approach / Jeff Barlow.
Description: First edition. | Burlington, MA : Jones & Bartlett Learning, [2019]
Identifiers: LCCN 2017055835 | ISBN 9781449641788 (paperback)
Subjects: LCSH: Command and control at fires. | Emergency management. | Decision making.
Classification: LCC TH9320.8 .B37 2019 | DDC 363.37/8068—dc23
LC record available at https://lccn.loc.gov/2017055835

6048

Printed in the United States of America
22 21 20 19 18 10 9 8 7 6 5 4 3 2 1

Dedication

1. To my Lord Jesus Christ who gave me the gift of leadership and the calling to the fire service; I am your humble servant.

2. To my wife, Pam, and my sons, Alex and Ryan, who sacrifice daily when duty calls me away; I love you.

3. To my parents, Linda and Randy, who taught my sister Kathy and me how to creatively and respectfully relate to people of any background to meet their needs and wants with integrity; I love you and thank you.

4. To my fire service brothers and sisters from Union, College Park, Eastern Kentucky University, Toyota, Bloomington, and Burlington; you have taught me so much and graciously allowed me to serve with you. I thank you.

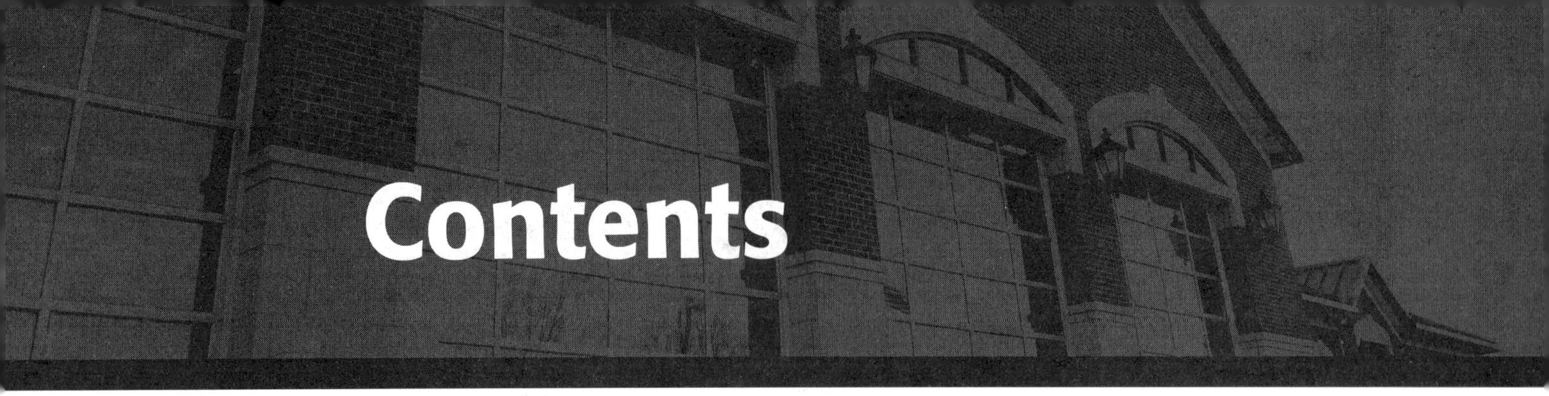

Contents

Section 1 Company Officer 1

Chapter 1 Company Officer Scenarios 3

Preface

Introduction

The purpose of this book is to assist fire officers in developing and fine-tuning decision-making skills. The fire service has emphasized the importance of emergency response size-up: a proper size-up and subsequent actions can make the difference between a successful operation and an operation riddled with risk, including injury or death. Training texts, seminars, and academies teach officer skills, but such training is often general in nature and focuses on fireground responsibilities. Rarely, if ever, does an officer receive training and practice in sizing up station and personnel scenarios and putting appropriate decisions into action. These situations involve the public, the officer's crew, and the other members of the department, all of which can influence the culture of the entire fire department. Even though the vast majority of situations for which an officer is solely responsible are in the station, an officer frequently relies on on-the-job training for guidance as he or she gains rank and is promoted up the chain of command.

In order to prepare aspiring officers to take on the challenges of their position before they are officially tasked with handling station issues, we must allow them to practice their decision-making skills. Most fire departments provide practice during the probationary period. The problem with this method is that departments rarely have systems in place to retract a promotion to an officer's position if someone exhibits a less than stellar performance as an officer during the probationary period. More often, the underperforming officer is allowed to keep practicing until he or she gets it right. This can have extensive organizational consequences that can discourage future leaders, detract from morale, and solidify a culture of frustration. It can also present a significant risk of liability to the entire organization.

An alternate school of thought is that, if someone does well in his or her current position and he or she tests or completes an assessment center well, then he or she will make a reasonably good officer when promoted. This makes sense in theory, but two important facts remain unaddressed.

First, when someone does well in his or her current assignment, while that positive performance is worthy of recognition or reward, it does not mean that person will necessarily do well in a new, more complex assignment.

Second, an assessment center, test, or promotion process generally involves only a handful of situations to be evaluated for "correct" or "incorrect" responses, and frequently significant emphasis is placed on emergency response–related judgments. Can a few brief questions during a promotion process reveal how well a person is prepared to handle those situations?

To truly practice sizing up station scenarios, the aspiring officer must approach situations from an officer's point of view with an officer's responsibilities, information, and expectations. The scenarios presented in this text allow aspiring officers to critically assess various station issues so that they are prepared to make sound judgment calls that result in the best possible outcomes when they face similar station issues as fire officers.

Words of Wisdom

Let experience continue to lead you on the road to more effective leadership. Just as you would not use a handful of prior scene size-ups to blindly dictate your specific actions at the next fire scene, do not let a few case studies definitively dictate how you will handle future challenges.

Policy and Procedure Limitations

The National Fire Protection Association's 1021 Standard, *Fire Officer Professional Qualifications*, states in section 4.2.5, "Apply human resource policies and procedures, given an administrative situation requiring action, so that policies and procedures are followed." The issue you will find with many case studies in this book is that most fire departments do not have policy or procedure guidance for specific human resource issues. There may be some general "do this" or "don't do that" human resource policies, such as you are not allowed to create a hostile work environment by making threats to co-workers or the public, and you are to report on-the-job injuries within 24 hours of the injury, but these will not help you to figure out how to handle specific situations. As a fire officer, you must make an educated, skillful judgment call based upon sound reasoning with minimal policy guidance.

Organization of the Text

This text is designed to identify situations at various levels of the organization and is divided into three parts: Company Officer, Battalion Chief/Shift Commander, and Operations Chief/Fire Chief/Chief Elected Official. Specific ranks are not always identified in the scenarios because agencies or departments may use various terminologies. For each level of responsibility within a department, new levels of complexity are added, consequences are magnified, and expectations and perspectives from above and below the officer's rank change. Some of the scenarios are repeated across the three levels. This is because officers at different levels, with different domains of responsibility, will size up the same situation differently. As an officer climbs the organizational chain of command, he or she will find that the same issue will require different approaches.

Scenario Format

Each scenario lists the applicable job performance requirements (JPRs) for NFPA 1021, *Standard for Fire Officer Professional Qualifications.*

Although the scenarios described in this book are fictitious, they are grounded in real-life experiences from a variety of sources, including industrial fire departments, volunteer fire departments, career fire departments, and combination fire departments. Some of the scenarios refer to volunteer members and career members. These references allow any type of department to use the information. If you are in an all volunteer department, simply read the scenario as all volunteer members; similarly, if you are in an all career department, read the scenario in that context. The issue and the challenge to the officer will remain the same.

To facilitate your understanding of how these various types of issues can impact other levels of the department, the case studies are grouped together by their issue classification. The case studies can be generally classified as:

- Issues of work product
- Interpersonal dynamics
- Sensitive policy matters
- Attitude issues

Many case studies are revisited as you progress through the book, enabling you to size up the case study from the perspective of a higher level fire officer, all the while considering the increased level of responsibility. This allows you to see how a fire officer's decision made in the station can have an impact throughout the entire fire department and even throughout the entire municipality.

Each scenario is followed by a series of thought-provoking questions to help guide you as you size up the given station issue. Discussion of the scenarios, including insights and sample responses to each of the scenario's questions, is included in separate chapters within the text. After you consider the questions and formulate your plan, you will be able to look in the next chapter to read discussion insights for each question to jump start your thought processes or compare to your responses. You are encouraged to write or type out your responses so that you can compare your approaches to the scenarios as you progress throughout the book. Since policies and regulations will vary across departments and jurisdictions, there may be several appropriate responses to a given scenario. The sample responses provided are intended to highlight key points.

These case studies are intended to be a safe environment to practice and assess your problem-solving skills. Take full advantage of this opportunity.

SECTION 1

Company Officer

Introduction

As an officer, you will find some people you supervise will have strong feelings about how you should make decisions. Often when two or more conflicting perspectives are brought to you to decide a course of action, the people with opposing perspectives will expect you to resolve the conflict in their favor. For example, Ted, a fire fighter on your crew, has been working hard to get the morning apparatus checks and station chores completed as soon as the shift begins. Sasha, another fire fighter on your crew, tends to enjoy a morning cup of coffee or two before starting on the checks and chores. After Ted gets tired of feeling like he is doing all the work while Sasha sips the morning away, he comes to you and wants you to do something about the situation. You know Ted expects you to hear his complaint, agree with him, and make Sasha do more of the work. You also know Sasha expects you to recognize she is willing to do the chores, agree with her, and tell Ted to slow down a bit and lighten up. In a perfect world you would pull Ted and Sasha together, talk about their perspectives, and everyone would agree on a course of action. Here is where the ideal and the real world often separate. If you decide to take any action to change Sasha's routine, she will perceive that you agreed with Ted's complaint. If you decide Sasha does enough work to meet expectations as it is, Ted will perceive you agreed with Sasha, and he might feel undervalued for his get-it-done attitude. If the concept of simply sitting Ted and Sasha down to talk about their concerns and agreeing to a solution was easy, Ted and Sasha would have done that on their own. In this situation, and many situations officers are faced with, the officer is the problem solver, the referee, and the mediator. This also means the officer easily becomes the "good guy" or the "bad guy" depending upon how he or she decides.

As if working through those two perspectives isn't difficult enough, your superior officers will expect you to decide matters effectively for the good of the organization. In the Ted and Sasha example, the battalion commander isn't concerned about who specifically does the checks and chores or how fast they get done. However, the battalion commander does expect you, as the company officer, to manage the company in such a way that

FIGURE 1-3 What if we move this over there and that over here?

Courtesy of Jeff Barlow.

4. Enthusiastic Fire Fighter Does Not Complete Station Assignments

▸ NFPA 1021: 4.2.2, 4.2.6, 5.2.1

Josh is a very enthusiastic fire fighter when it comes to sharing his knowledge with others, but he often fails to complete routine station assignments. When Josh receives your assignment to put up the station flag and wash the reserve engine, he typically starts talking to the less senior fire fighters on the apparatus floor about the features of the reserve engine or the quirks of the pump panel. This leads to an informal pump operations or hose-handling training session or even a dissection of department policy about response procedures. When walking through the apparatus bay 30 minutes later and passing by the watch office door, you observe the station flag still lying on the watch desk, and the reserve engine remains unwashed.

1. **If you do nothing, what will happen?**
2. **What do you say or do?**
3. **With whom do you address the issue?**
4. **What are two other alternatives you could use?**
5. **What is your last-resort action?**
6. **If you reach your last-resort action, what effect will you be dealing with two weeks after taking action?**

5. Instructor Teaches Incorrect Material

▸ NFPA 1021: 4.2.3, 4.2.6, 5.2.1, 5.2.2

Aaron is a senior fire fighter and certified instructor assigned to train others in the department. While sitting through one of his classes, you realize that he is teaching incorrect information. Today he is instructing a multicompany drill in which he emphasizes the importance of salvage and overhaul. He teaches responders to remove all contents of any structure fire down to the stud walls to ensure that the fire is overhauled properly and that fires do not rekindle. He quotes from published training texts about the importance of overhaul and adds that fire investigators need to understand the hard work of response companies. He assures the class that any fire investigation evidence can simply be rebuilt outside the structure for photo or evidence collection for a fraction of the effort fire fighters spend battling the blazes.

FIGURE 1-4

1. **If you do nothing, what will happen?**
2. **What do you say or do?**
3. **With whom do you address the issue?**
4. **What are two other alternatives you could use?**
5. **What is your last-resort action?**
6. **If you reach your last-resort action, what effect will you be dealing with two weeks after taking action?**

6. Fire Fighter Brings Non–Fire Department Work into the Station

▸ NFPA 1021: 4.2.2, 5.2.1

Chip is a good and effective fire fighter on your engine company, who works in construction as a general contractor on his days off. In the evenings during his shifts at the station, after dinner is finished and the kitchen is clean, Chip tends to spread out job plans and blueprints for the next day's contractor work. Everyone at the station likes Chip, but when he spreads out his documents, he prevents the other crew members from using the table to eat a snack, work on a fire department project, or even study fire training materials. This is not happening just once in a while—it occurs almost every night when Chip is working. You observe that Chip's plans take up almost all of the table, forcing others to go elsewhere or wait until Chip is finished, a few hours later.

1. **If you do nothing, what will happen?**
2. **What do you say or do?**
3. **With whom do you address the issue?**
4. **What are two other alternatives you could use?**
5. **What is your last-resort action?**
6. **If you reach your last-resort action, what effect will you be dealing with two weeks after taking action?**

7. Fire Fighter Engages in Non–Fire Department Work During Training and Responses

▸ NFPA 1021: 4.2.3, 4.2.6, 5.2.1

Marcus is a fire fighter on your engine company and also a general contractor for his own construction business, which he runs on his days off. He receives numerous phone calls throughout the day on his cell phone regarding his construction duties and subcontractors. He even receives some calls during training and on emergency responses. When he first joined your company months ago, Marcus would apologize for interrupting a training session, but now he simply excuses himself from the session and steps out of the training room or away from the hands-on activity to talk through the issue on the phone. The calls last for several minutes, and recently Marcus has been observed making outbound follow-up calls after the initial interrupting phone call. In addition, Marcus carries his phone with him on the engine because he says some calls about concrete or coordination are time sensitive. He even took one call on an active alarm response while the company was staged awaiting assignment.

FIGURE 1-5

1. **If you do nothing, what will happen?**
2. **What do you say or do?**
3. **With whom do you address the issue?**
4. **What are two other alternatives you could use?**
5. **What is your last-resort action?**
6. **If you reach your last-resort action, what effect will you be dealing with two weeks after taking action?**

8. Fire Fighter Requests Training to Perform Duties Not Applicable to the Department

▸ NFPA 1021: 4.2.3, 5.2.3

Mary submits to you a written training request form that asks permission to take advanced wildland fire training classes to qualify her for deployment on large western wildfires. This specialty fire training is not applicable to your department's urban/suburban jurisdiction. In the kitchen later that same day, you hear some fellow crew members mention that Mary is preparing to leave this job and find work elsewhere. As you walk back to your office, you consider how to assist Mary appropriately while still acting in the best interest of your department.

1. **If you do nothing, what will happen?**
2. **What do you say or do?**
3. **With whom do you address the issue?**
4. **What are two other alternatives you could use?**

5. **What is your last-resort action?**
6. **If you reach your last-resort action, what effect will you be dealing with two weeks after taking action?**

11. Career Fire Fighter Questions Volunteer Fire Fighters' Skills

▸ NFPA 1021: 4.2.3, 4.2.4, 4.2.5, 5.2.1

While sitting in your office late one night after a training session, you are approached by Frank, a long time volunteer member. Frank tells you that you have to do something about Tony, a recently hired career member. Frank explains that Tony "struts" through the bay with the other volunteers in attendance for training and is often heard telling the volunteers that if they work hard, they, too, can someday be hired and learn how to be a "real" fire fighter and not just a "play fire fighter" when a call comes in. Frank perceives that Tony is arrogant and condescending and is concerned that Tony's attitude is negatively affecting the company.

1. **If you do nothing, what will happen?**
2. **What do you say or do?**
3. **With whom do you address the issue?**
4. **What are two other alternatives you could use?**
5. **What is your last-resort action?**
6. **If you reach your last-resort action, what effect will you be dealing with two weeks after taking action?**

12. Conflict Between Two Fire Fighters on Different Shifts

▸ NFPA 1021: 4.2.2, 4.2.4, 4.2.5, 5.2.1

Two members of your crew—John, an apparatus operator for your rescue company, and Sarah, a fire fighter on a different shift but in the same station—are in personal conflict. John and Sarah have an oil-and-water relationship: Everything Sarah says seems to grate on John's nerves, and vice versa. Although John's and Sarah's interactions are mostly limited to those occasions when they pass each other during the shift change every third morning or when a rare overtime shift or shift trade occurs, these moments are tense. Even the shift trades are infrequent because neither John nor Sarah really wants to spend a shift together if it can be prevented. When they do speak to each other, the tone is pointed and sarcastic, and the exchanges are contentious. Light conversation about food or cars ends in discord. What one likes, the other seems to hate. As the officer, you have observed this negativity between your two crew members first hand, but other rescue company crew members have also come to you to express their concerns. Although John and Sarah typically interact only during a brief shift change procedure, you are keenly aware that the ensuing tension sets an unwanted tone for your crew for the day.

1. **If you do nothing, what will happen?**
2. **What do you say or do?**

3. **With whom do you address the issue?**
4. **What are two other alternatives you could use?**
5. **What is your last-resort action?**
6. **If you reach your last-resort action, what effect will you be dealing with two weeks after taking action?**

13. Conflict Between Two Fire Fighters in the Same Company

▸ NFPA 1021: 4.2.4, 4.2.5, 5.2.1

Stan is a member of your rescue company and has been an active fire fighter and rescue technician for more than five years. Stan has a personal conflict with Jake, another member of your rescue company. Jake is the apparatus operator for the rescue company and has been under your supervision for 11 months. Stan and Jake simply do not get along. When not in your presence, they speak to each other with a nasty and sarcastic tone. This negativity especially tends to surface whenever they must work together to handle nonemergency business such as cooking, training, or station chores. They can perform assigned tasks effectively but offer no voluntary assistance to each other throughout the shift. On a personal level, they disagree on politics, religion, and many other topics that come up in casual everyday conversation with the other rescue company members. You are aware of the sarcastic comments and tense work atmosphere every shift day, all day long. The personal conflict is affecting the entire rescue company by creating a biting and negative atmosphere during the company's workday. Positive events and encouraging feedback from outside the company are often overshadowed by the constant conflict.

1. **If you do nothing, what will happen?**
2. **What do you say or do?**
3. **With whom do you address the issue?**
4. **What are two other alternatives you could use?**
5. **What is your last-resort action?**
6. **If you reach your last-resort action, what effect will you be dealing with two weeks after taking action?**

14. Fire Fighter Disregards Chain of Command

▸ NFPA 1021: 4.2.2, 4.2.5, 4.4.4, 5.2.1

Bill is a two-year veteran fire fighter of the department and a recent transfer to your engine company. Whenever he has operational or administrative questions, Bill routinely bypasses you as the company officer and solicits answers from the chief officers instead. It is not unusual to walk by the chief's office or the assistant chiefs' offices during the course of any day and find Bill standing in the doorway inquiring about why the department purchases apparatus soap from a particular vendor or which content might be on an upcoming promotion process so he can be well prepared. Bill never brings these types of questions to you, his immediate superior. Sometimes Bill even reports the questions' answers back to you as if to keep you informed. You are frustrated with Bill's ignoring the chain of command.

FIGURE 1-8

1. **If you do nothing, what will happen?**
2. **What do you say or do?**
3. **With whom do you address the issue?**
4. **What are two other alternatives you could use?**
5. **What is your last-resort action?**
6. **If you reach your last-resort action, what effect will you be dealing with two weeks after taking action?**

15. Fire Fighter Fraternizes with City Manager

▸ NFPA 1021: 4.2.5, 4.4.4

Your department has a chain of command that all members understand and are expected to follow. You learn through passing comments of other fire fighters around the station that Kevin, a member of your ladder company, has become increasingly friendly with the city manager. Kevin did not previously know the city manager, but since Kevin was elected as the fire fighters' labor representative, the city manager has extended various dinner invitations to Kevin and given him tickets to sporting events. In some instances, the only city personnel in attendance were the city manager and Kevin. The city manager's family and Kevin's family have also attended multiple family picnics together. While casually chatting in the station kitchen, Kevin mentions that fire department business sometimes enters his off-duty conversations with the city manager.

1. **If you do nothing, what will happen?**
2. **What do you say or do?**
3. **With whom do you address the issue?**
4. **What are two other alternatives you could use?**
5. **What is your last-resort action?**
6. **If you reach your last-resort action, what effect will you be dealing with two weeks after taking action?**

Sensitive Policy Issues

16. Last-Minute Arrival for Shift Changes

▸ NFPA 1021: 4.2.4, 4.2.5, 4.2.6, 5.2.1

As the station lieutenant, you try to set a good example for your crew by being in the station at least 15 minutes before shift change to ensure you can be ready for work the moment the clock strikes shift change time. Ian, one of your senior fire fighters, usually does not arrive at the station until immediately before the shift change. Ian is often seen hastily slipping on and buttoning up his uniform shirt as he runs across the parking lot to enter the station just as the clock strikes the 8:00 A.M. shift change. Kirk, the off-going fire fighter in Ian's riding position, understands that shift change does not occur until 8:00 A.M., but he also knows that when the company is called for an emergency minutes before shift change, he will be responsible for responding because Ian is unlikely to be on the fire department property. Kirk, like most shift fire fighters, understands that things happen and help is needed every once in a while, whether because of a big fire response late in the shift, a fire fighter's sick child, or a flat tire on the way to work. Kirk has no issue with helping another fire fighter on occasion, but Ian's just-in-time arrival has become a habit. This pattern creates scheduling hardships for Kirk and results in overtime for the department. Kirk is discouraged by what he feels is Ian's lack of respect for him, and he looks to you to address the situation.

1. **If you do nothing, what will happen?**
2. **What do you say or do?**
3. **With whom do you address the issue?**
4. **What are two other alternatives you could use?**
5. **What is your last-resort action?**
6. **If you reach your last-resort action, what effect will you be dealing with two weeks after taking action?**

FIGURE 1-9 Being at work on time and being ready for work on time are two different things.
Courtesy of Jeff Barlow.

17. Fire Fighter Abuses Sick Time Policy

▸ NFPA 1021: 4.2.5

Jeff, a crew member on your ladder company, calls in sick for his 24-hour shift. Jeff follows department procedure and notifies you and the shift scheduler two hours before shift change. Later that day, while you and the ladder company are performing commercial fire prevention inspections, you notice Jeff's pick-up truck at the bowling alley across the street from where you are conducting inspections. You are able to identify Jeff's truck because of the fire department sticker on the window and the custom graphics he installed on the tailgate. While conducting inspections, one of your fire fighters sees and alerts you to Jeff exiting the bowling alley with some other adults and children. He is holding a soft drink can and does not appear to be sick.

1. **If you do nothing, what will happen?**
2. **What do you say or do?**
3. **With whom do you address the issue?**
4. **What are two other alternatives you could use?**
5. **What is your last-resort action?**
6. **If you reach your last-resort action, what effect will you be dealing with two weeks after taking action?**

18. Fire Fighter Smells of Alcohol

▸ NFPA 1021: 4.2.4, 4.2.5

Rita, a member of your station ambulance crew, has been with the department for more than five years. She has previously required performance coaching sessions from officers for missed chores, minor attendance issues, and similar infractions, but there is nothing problematic in her permanent employee file. Around the station, Rita is known as an easy spirit who enjoys socializing both with other off-duty fire/EMS personnel and nonfire/EMS friends and acquaintances. This morning when Rita entered the station kitchen, some crew members noticed the smell of alcohol. One of the crew members, Bennett, a fire fighter temporarily assigned to your fire engine for the shift, sticks his head through your office door and says, "Lieutenant, you might want to check out the kitchen. Something, or some*one*, seems to smell over there."

1. **If you do nothing, what will happen?**
2. **What do you say or do?**
3. **With whom do you address the issue?**
4. **What are two other alternatives you could use?**
5. **What is your last-resort action?**
6. **If you reach your last-resort action, what effect will you be dealing with two weeks after taking action?**

19. Fire Fighter Under the Influence

▸ NFPA 1021: 4.2.5

José, a fire fighter from an engine company across town, has been detailed to your company for the next six weeks while one of your regularly assigned fire fighters is recovering from an injury. José has a reputation for both working hard and playing hard. He has been with the department for more than 10 years and has not had any disciplinary issues of which you are aware. José is rumored to have been out partying the night before, based on Facebook posts that other members of your crew are discussing. He reports for duty and disappears by moving to a remote area within the station. Any time other crew members approach where José is working, he moves to another remote location in the station. He tries to escape to his bunk room as much as possible. As the officer, you notice that station duties do not include significant amounts of time in the bunk room, and your crew is starting to talk in your presence about how they have not seen much of José.

1. **If you do nothing, what will happen?**
2. **What do you say or do?**
3. **With whom do you address the issue?**
4. **What are two other alternatives you could use?**
5. **What is your last-resort action?**
6. **If you reach your last-resort action, what effect will you be dealing with two weeks after taking action?**

24. Fire Fighter Self-Discloses an Arrest

▸ NFPA 1021: 4.2.4, 4.2.5

It is shift change on a Saturday morning. One of your crew members, Brian, a fire fighter and move-up apparatus operator, approaches you. Brian says that he was arrested for an alleged assault the prior Thursday. He posted bond and was released, but he will be making court appearances soon. Brian tells you that the allegations are false; he says that his girlfriend called the police and conjured up false claims about him assaulting her. Brian tells you he needs to be at work for the income and the distraction from the current events.

FIGURE 1-10

1. **If you do nothing, what will happen?**
2. **What do you say or do?**
3. **With whom do you address the issue?**
4. **What are two other alternatives you could use?**
5. **What is your last-resort action?**
6. **If you reach your last-resort action, what effect will you be dealing with two weeks after taking action?**

25. Driver/Operator with Suspended Driver's License

▸ NFPA 1021: 4.2.2, 4.2.4, 4.2.5

Tim, a fire fighter in your crew, approaches you during morning chores while you are both outside behind the station. Tim tells you that he overheard Steve, the fill-in driver/operator who is driving the fire apparatus today, say he has a suspended driver's license. While he was checking his breathing apparatus at the start of the shift, Tim heard Steve laughing with a couple of fire fighters going off-shift about the fact that eventually he would have to obtain a hardship exception for his suspended driver's license. Steve also boasted that as long as he was not pulled over, he would be fine. Tim did not agree with that view, and he wanted to make sure you are aware of the situation, since Steve is detailed to your company to drive for the shift.

1. **If you do nothing, what will happen?**
2. **What do you say or do?**
3. **With whom do you address the issue?**
4. **What are two other alternatives you could use?**
5. **What is your last-resort action?**
6. **If you reach your last-resort action, what effect will you be dealing with two weeks after taking action?**

26. Fire Fighter Violates No-Tobacco-Use Policy

▸ NFPA 1021: 4.2.5

Your department has a strict no-tobacco-use policy. In the policy, smokeless tobacco and smoking tobacco are treated the same. One of your crew members reported that someone on your crew is sneaking chewing-type tobacco into the building and is chewing discreetly in violation of the policy. The crew member making the report wants to remain anonymous and refuses to identify the culprit.

1. **If you do nothing, what will happen?**
2. **What do you say or do?**
3. **With whom do you address the issue?**
4. **What are two other alternatives you could use?**
5. **What is your last-resort action?**
6. **If you reach your last-resort action, what effect will you be dealing with two weeks after taking action?**

Attitude Issues

27. Fire Fighters Critical of Fire Department Decisions

▸ NFPA 1021: 4.2.2, 5.2.1

You are a member of a four-person ladder company. Roger is your senior fire fighter, and Brett is your apparatus operator. Roger and Brett complain constantly about almost every decision made in the company and the department. They grumble during morning house chores, during meals, and while returning from calls or assigned work details. Sometimes they complain about company activities or priorities of work assignments. They express displeasure with department decisions, such as changes in response areas or company assignments at active fire alarm responses. Brett and Roger seem to feel that they have a better approach to every department issue and decision, and they are quick to comment when they think a decision, plan, or activity is "stupid." They tend to make their comments to each other, usually within earshot of others in the firehouse, so they cannot be corrected for direct insubordination. When any crew member offers support or positive reinforcement regarding the company or department, Brett and Roger scold that crew member for his or her failure to recognize senseless decisions. Brett and Roger appear to be entertained by their negativity and seem to enjoy expressing their dislike for the department overall, stemming from the decisions of the company, shift, and chief leadership. Their negativity overshadows any positive thoughts or deeds, and it is depressing and frustrating the other crew members.

1. **If you do nothing, what will happen?**
2. **What do you say or do?**
3. **With whom do you address the issue?**
4. **What are two other alternatives you could use?**
5. **What is your last-resort action?**
6. **If you reach your last resort-action, what effect will you be dealing with two weeks after taking action?**

28. Volunteer Fire Fighters Compare Incentive Stipend Checks

▸ NFPA 1021: 4.2.4, 4.2.5

Volunteer members of your department receive stipend checks as incentives to participate in firefighting activities. Check amounts are based on levels of activity and training over a period of time, so members receive different amounts depending on their individual participation. Emma has been a volunteer fire fighter with your department for two years and recently received a volunteer incentive stipend check. While talking with Seth, a volunteer fire fighter who has been with the department for seven years, after a training one evening, Emma learned that Seth received a larger incentive check. Emma states in the presence of you and other crew members that she feels cheated and unappreciated. Although Seth is an involved member, Emma believes she is more active than he is, attending training at the fire station almost every week and going on many of the department's response calls. Seth is satisfied with his current check amount and does not want anything to change. As far as you are aware, no other volunteers are comparing incentive stipend checks.

FIGURE 1-11 This can't be right. I know I have been around here more than this shows!
Courtesy of Jeff Barlow.

1. **If you do nothing, what will happen?**
2. **What do you say or do?**
3. **With whom do you address the issue?**
4. **What are two other alternatives you could use?**
5. **What is your last-resort action?**
6. **If you reach your last-resort action, what effect will you be dealing with two weeks after taking action?**

29. Fire Fighter Critical of Pay Grade System

▸ NFPA 1021: 4.2.4, 4.2.5, 5.2.3

Juan, a crew member in your water tender company, is considering a promotion opportunity. Based on the published job announcement and the current department policy related to pay rates, he compares the pay rate for the next rank up, sergeant, against his current pay rate

FIGURE 1-12

1. **If you do nothing, what will happen?**
2. **What do you say or do?**
3. **With whom do you address the issue?**
4. **What are two other alternatives you could use?**
5. **What is your last-resort action?**
6. **If you reach your last-resort action, what effect will you be dealing with two weeks after taking action?**

33. Company Dissatisfied with Design of New Apparatus

▸ NFPA 1021: 5.2.1, 5.4.6

You are the company officer for Engine Company 14. A new piece of fire apparatus is assigned to your station. Your crew members do not like the compartment design of the newly received fire apparatus. They think the storage compartment located inside the cab is too big, reducing the room available for the crew, and is poorly designed, because the equipment intended to go inside that compartment could be stored more conveniently in other places. The engine is the first brand-new apparatus the station has received in years, and you believe that your crew is unappreciative. They keep finding what they perceive to be problems, such as equipment shelf locations in the apparatus body compartments, the location of switch panels for the apparatus operator, and the rear-facing seating design. They blame the seat design on the large storage compartment in the cab, which they believe is unnecessary.

1. **If you do nothing, what will happen?**
2. **What do you say or do?**
3. **With whom do you address the issue?**
4. **What are two other alternatives you could use?**
5. **What is your last resort-action?**
6. **If you reach your last-resort action, what effect will you be dealing with two weeks after taking action?**

34. Excessive Equipment Location Changes Request

▸ NFPA 1021: 4.2.6, 5.2.1, 5.4.6

Another station in your department has received a brand-new quint fire apparatus. Due to your station's relatively slower activity level, the quint fire apparatus that the other station had been using is reassigned to your station to finish out its life cycle within the department. The quint is still in fair condition, but once it arrives at your station, your crew begins looking for ways to change equipment to make it more user friendly. Tara, a two-year fire fighter, comes to you almost every shift with a new idea for where to move various hand tools and how to rearrange the fans and saws. At first, you approved a few of the proposed moves, but Tara keeps coming to you about once a week with a new idea. First, she wanted to move the extinguishers out of the compartment and mount them on the side running boards. Next, she suggested moving the apparatus operator's turnout equipment from the first compartment to the rear compartment. Then, she advocated adding a standpipe hose

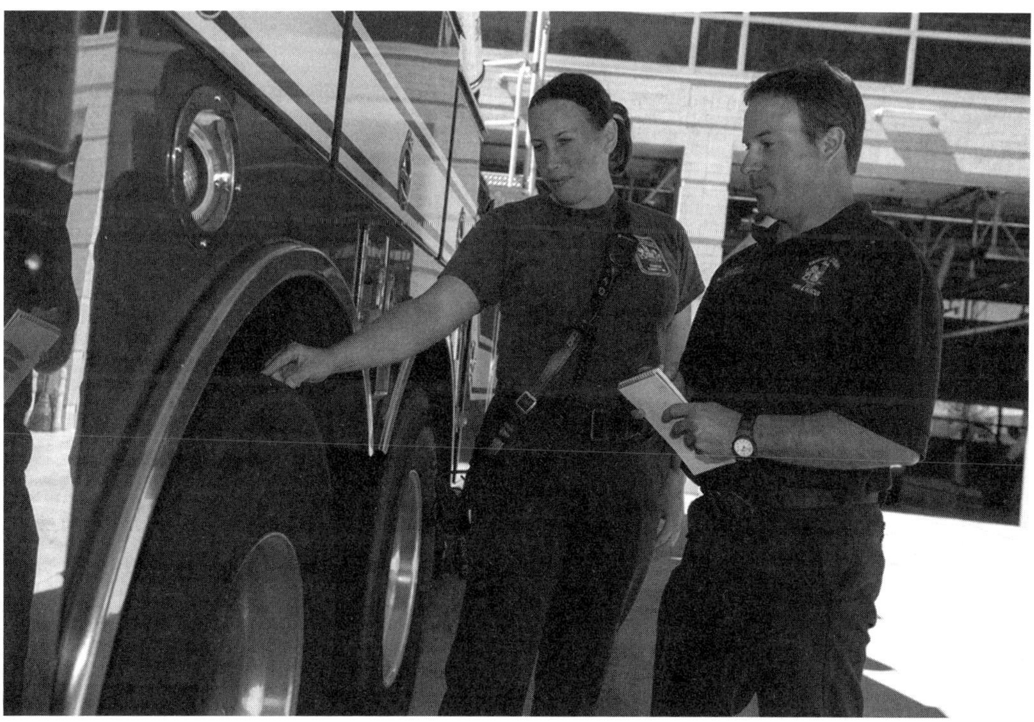

FIGURE 1-13

pack to the officer's side compartment, followed three weeks later by a request to move the standpipe hose pack to the driver's-side compartment because it would work better. You have heard from some of your fire fighters that your colleagues on the shift before and the shift after yours are complaining to their officers about the frequent equipment relocations. The other shifts are becoming frustrated because those personnel never seem to know where the equipment is located when they need it.

1. **If you do nothing, what will happen?**
2. **What do you say or do?**
3. **With whom do you address the issue?**
4. **What are two other alternatives you could use?**
5. **What is your last-resort action?**
6. **If you reach your last-resort action, what effect will you be dealing with two weeks after taking action?**

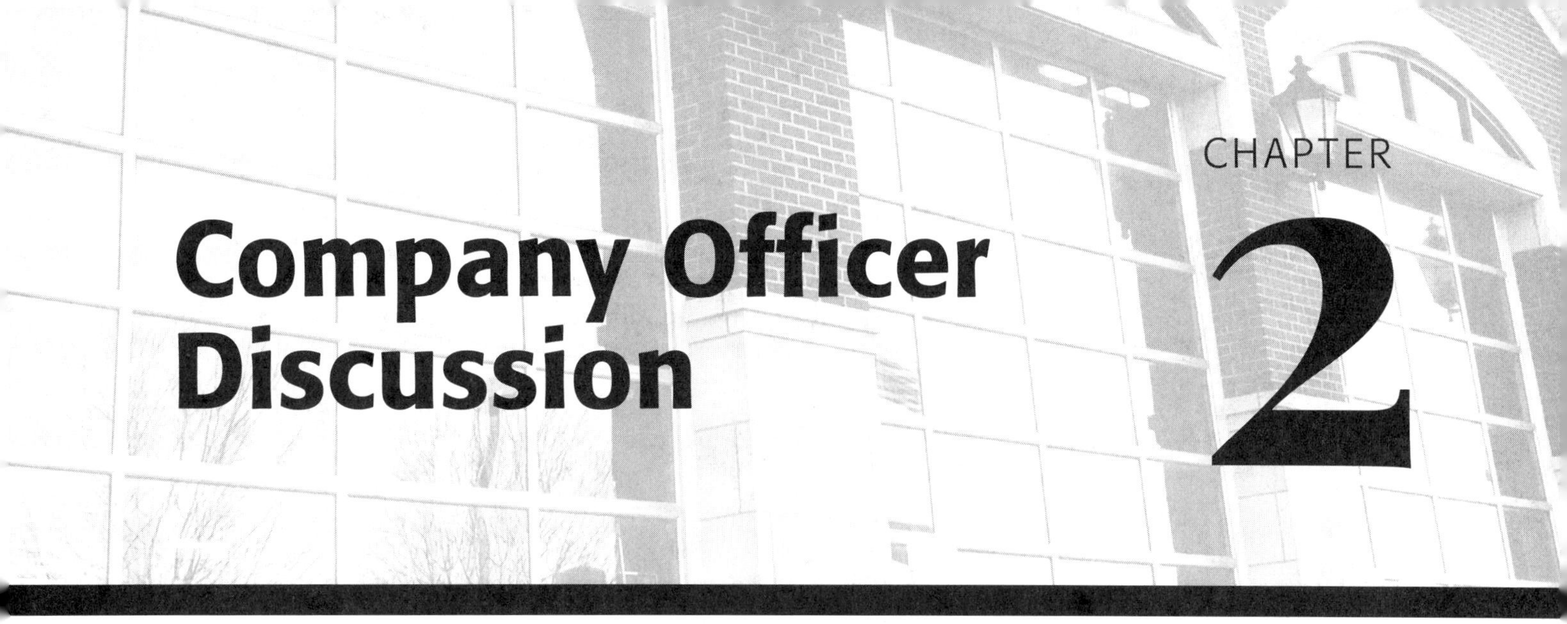

CHAPTER 2

Company Officer Discussion

Work Product Issues

1. Dirty Dishes

▸ NFPA 1021: 4.2.2, 4.2.6

Your department and your station expect that every individual will clean up after himself or herself. Chore lists, supervision, and history all reinforce the assumption that the fire station kitchen, along with the rest of the station, is to be kept clean and orderly. When you report for duty at the station, you typically start your shift by going to the kitchen for a cup of coffee. Today you discover that the previous shift did not clean up the station kitchen as expected. The floors are not swept, the table is dirty, and the sink contains unwashed dishes. You have previously noted an occasional oversight of a dirty coffee cup left behind for the oncoming shift, but recently the volume and frequency of items left behind have increased, and today the kitchen will require some effort to clean up. Since the previous shift has gone home, it is now your shift's responsibility to clean up the mess.

1. If you do nothing, what will happen?

If the issue is not addressed, it will only escalate. This situation has the potential to create animosity toward an individual or a shift or crew in general. Crews may start to complain that they are unfairly burdened with kitchen-cleaning duties or simply refuse to clean up after other crews. Either way, this issue creates a distraction at a minimum and more likely contention between the crews. As the company officer, you must act.

2. What do you say or do?

First, let your crew know that you are aware of the problem and that you are taking steps to address it. Encourage crew members not to take matters into their own hands or to allow this issue to upset them. Assure them that as their company officer, you will handle this situation. This will let your crew know that you hear their concerns and give you the time to address the issue with the previous shift's crew.

Next, you should have a discreet but direct conversation with your peer officer from the previous shift. During the conversation, do not make assumptions or accusations. As one officer to another, simply explain the facts as you witnessed them: This is what was found, this is when it was found, this is how often it occurs, and this is where it occurs. Share your concerns and the concerns of your crew with your peer. Ensure that the company officer understands that you are bringing this issue directly to his attention in an effort to have it corrected quietly and without emotion. Your goal is to ensure that both crews can refocus their attention on what is important: responses, training, inspections, and other shift responsibilities.

3. With whom do you address the issue?

First, speak with your crew. Next, speak with your peer-level officer. Finally, inform your crew that the issue is being addressed. There is no need to share specifics from your conversation with the other company officer, but assure your crew that the conversation with the appropriate personnel has taken place.

4. What are two other alternatives you could use?

If you know who is responsible, you could pull that individual aside one day during the shift change and speak directly with him or her. If you take this course of action, use the same positive tone that you would have used with your peer officer. Reassure the fire fighter that this issue needs to be addressed to keep each crew focused on firefighting and not dirty dishes. Immediately follow this conversation with a conversation with the fire fighter's officer as a peer courtesy. You can explain to the fire fighter's officer that you handled the issue directly because you are both officers and share the same duties. Differing shift assignments do not mean that a fire fighter is "off limits" from another officer's correction; however, discussing the correction with the peer officer as a professional courtesy is a must.

A second option might be to raise the issue with the other officer's supervisor. However, asking an officer's supervisor to get involved as the first step in conflict resolution is dangerous for peer relations. It would also signal a lack of ability to effectively manage a minor station issue as a company officer. This could erode hard-won credibility.

5. What is your last-resort action?

Your last-resort action will be influenced by many variables, based on your department's culture, regulations, and attitudes. In one department, the last resort may be to issue a group written warning for the entire shift; another department's culture may require you to bring in the shift commander to assist with resolution of the issue; in another department, the solution may be to pile all dirty dishes in a bin until there are no more clean dishes. In such cases, the last resort may be to use disposable plates, cups, and utensils instead.

6. If you reach your last-resort action, what effect will you be dealing with two weeks after taking action?

Reaching a last-resort action over dirty dishes may indicate a deeply rooted department culture challenge that might require outside assistance from superior officers or possibly even outside organizational leadership consultants. Two weeks after implementing the last-resort action, you will likely be dealing with disgruntled fire fighters who are dissatisfied with your decision. There is neither a fast nor an easy solution to addressing these subsequent effects. The key for you as the company officer is to neither give up nor become discouraged. You are in a position to create a positive influence, and you may need to develop a strong sense of patience as you lead in this type of department culture.

2. Low Fuel Gauge

▸ NFPA 1021: 4.2.2, 4.2.6

You are starting your shift and are walking to your office to assign tasks for the day. As you pass through the apparatus bay, your crew member John comes to you and reports that the fire apparatus is not being fueled regularly. As the regularly assigned apparatus operator at your engine company, John performs a quick check of the pumper at the start of every shift. He tells you he is fed up with being left to deal with a low fuel level, shift after shift.

1. If you do nothing, what will happen?

For most companies, unless they are located in a very busy urban area, nothing would happen. If the regulation says fuel at three-fourths full and the apparatus makes it to half full, the apparatus will still make the calls needed almost all of the time until it can be filled by John, your attentive apparatus operator who reported the problem to you. Even if your company is dispatched on "the big one," you will probably need a fuel truck on site to refuel other apparatus as well. This does not mean the issue should not be addressed, but rather you as the company officer must keep reality in perspective. When your apparatus operator comes to you with this issue, he is looking for heads to roll. Your crew is convinced that the next call, even before you can get the apparatus refueled, will be a six-hour conflagration with lives at risk and you will run out of fuel on the way just two blocks short of your destination. Remember, a good officer can see through the smoke and not overreact. Size up the issue and see if it is a symptom of a bigger problem or if it is just a fuel issue, and handle it appropriately.

2. What do you say or do?

Acknowledge the issue with your crew member, John. You may make the decision to tell your crew to forget about it, with your crew just taking on the responsibility to fuel the apparatus. It is unlikely a previous shift is intentionally trying to park the apparatus with less than appropriate levels of fuel. Just fuel it. Alternatively, you might decide this is an infraction from the preceding shift that compromises response capability. If that seems to be the case, address the issue with your peer officer on the preceding shift. Don't accuse, don't whine, don't look for someone to blame to satisfy your apparatus operator—just calmly talk through the issue and the implications, and then let the other officer have time to address it with his or her crew.

3. With whom do you address the issue?

Talk with John, your apparatus operator, first; then discuss the issue with the other officer if necessary. There is no need to address the issue with anyone else (unless you rotate apparatus operators and another apparatus operator in your crew has the same issue).

4. What are two other alternatives you could use?

Leave the apparatus unfueled and hope you do not leave the station; then someone else can deal with the problem—except now you are part of the problem, not the solution! Remember, you are the company officer and have a responsibility to fix any errors or omissions you discover, regardless of their cause. As a second option, you could talk with the preceding apparatus operator directly, but be warned: This can create a volatile situation. This issue is small enough that it will likely be perceived as sniveling because you, the company officer, are addressing it directly with the apparatus operator. The most professional course of action would be to go through the preceding company officer if you make the decision to take it outside your crew at all.

5. What is your last-resort action?

If the issue becomes a failure to complete an assigned task by the apparatus operator, you might have to look at counseling the individual apparatus operator. This counseling would not occur because the other shift complained but because your expectations were not met. If the issue is about the other shift not fueling the apparatus, your last-resort action may be to address the concern with a peer-level officer in an uninvolved station to see if that individual can mediate the situation. Avoid taking a fueling issue up the chain unless you have exhausted absolutely every option available at your rank/assignment.

6. If you reach your last-resort action, what effect will you be dealing with two weeks after taking action?

You will likely have growing frustration from both shifts—but do not give up and ignore the issue. It will not change on its own, and as the company officer, it is your responsibility to deal with these problematic issues.

3. Excessive Suggestions for Operational Changes

▸ NFPA 1021: 4.2.2, 4.2.6, 5.2.1

Sam is a fire fighter with many good ideas and is always offering suggestions for changing equipment locations on the ladder truck. Every time a new tool comes to market, he visualizes a use for it and comes to you for permission to rearrange existing equipment to make space for the new tool. These requests have come to you about twice a month, every month, ever since Sam was assigned to your company a few years ago. Once space is available in the equipment compartment, Sam asks you to act on getting the desired tool approved for purchase. For every request, Sam has a compelling argument for the purchase. First, it was a longer hook tool, followed by a different folding ladder, and then, six months ago, replacement of all the pick-head axes with flat-head axes. This afternoon, Sam comes to your office and asks you to reverse the prior decision and grant permission to remove the flat-head axes and replace them with pick-head axes.

1. If you do nothing, what will happen?

One of three things is likely to occur. First, Sam may eventually stop bringing you any ideas and will become frustrated. Second, Sam may find someone else to ask, and it could cause that person to wonder why you are not dealing with these requests from your crew member. Third, Sam might make the changes on his own and await any consequences. A freelancer is born!

2. What do you say or do?

First, hear Sam out. Listen to see if his ideas are something that might improve the operation. Next, if the ideas are actually plausible, delegate them. This is where you can impact not just your leadership skills but also those of your crew. Have Sam write his proposal succinctly and clearly for review with the other personnel affected, other shifts, other volunteers, and so on. Include a signature blank for agreement or dissent, and then let Sam run with the idea. Give him a deadline to report his findings and recommendations back to you. Offer to assist if needed by sharing the information with peer-level officers that your fire fighter is researching this project and you would appreciate other crews participating.

3. With whom do you address the issue?

Have a conversation with Sam. People involved in the fire service often communicate well informally—conversations about fishing, hunting, or sports come easily—but many lack the skill of clearly discussing real issues (not assigning or delegating but participating in an exchange of ideas) one-on-one and face-to-face. You as the officer must take the initiative to have the discussion about the reasons why Sam feels the suggested reorganization is better than the current layout. Challenge him to convince you with facts, not his opinions, and charge him with finding valid reasons why the equipment is currently where it is. Sometimes these constant suggestions can become an informal research project, causing the officer to research the issue to defend the current placement or find enough data from other sources to be convinced to change. Put this research squarely where it belongs—with the person making the suggestion.

4. What are two other alternatives you could use?

Give Sam the latitude for a trial period. This may be a trial period only on your crew/shift in the beginning. If that is the case, make sure equipment is returned to its "start" position when the next shift reports for duty. Another alternative might be to assign Sam to formally write up his proposal, including its benefits, potential drawbacks, and costs. Do not make this documentation alternative seem to be a deterrent for Sam. We have all been in meetings or groups where everyone seems hesitant to raise a concern or ask a question because they fear they will be tasked with running the subsequent project. Use this option when you need the member to transition from making verbal suggestions to becoming a change agent for the department by focusing on root issues and clarifying ideas to be shared with others who are not present.

5. What is your last-resort action?

If the suggestions are coming in fast and furious, and even with research assignments you continue to face a barrage, a last-resort action might be to tell Sam, "Enough, for now." Explain that any organization can implement only so many changes at a time. Reassure him that this is a temporary hold, and follow up in a reasonable period of time to jump-start the suggestion process.

If it is your workload that is the determining factor, explain that constraint to Sam. You may have an opportunity to assign other responsibilities to Sam and let the equipment placement drop on the priority list for *both* of you.

6. If you reach your last-resort action, what effect will you be dealing with two weeks after taking action?

The effects will hinge largely on how well the dialogue between you and Sam proceeded. If you were able to convey your sincerity and your genuine desire to consider these suggestions at a later point, things will likely be fine. In fact, your situation may improve because Sam feels included and may be helping you with other priorities.

If the "discussion" was more about both sides telling their story and expecting the other point of view to come around, then frustration will likely build in the station. If you see the meeting between you and Sam to be a necessary evil as part of moving on, Sam will sense that attitude, and the ugly seed of discontent will be planted and begin to grow. Sam will feel put off or, even worse, might feel that his input was not considered at all. You may become frustrated that Sam's attitude is declining. It is possible that you may start to feel Sam is questioning your authority to make this decision and move this issue up the chain of command. Potentially, Sam might start spreading the sentiment that you just do not care about his ideas. This is the beginning of a downward spiral that will negatively impact you, the crew, and even the department, and the whole matter will have started with Sam's perceptions, not the facts. The lesson here is to take the time to genuinely discuss all perspectives and reach understanding. It is worth every second!

4. Enthusiastic Fire Fighter Does Not Complete Station Assignments

▸ NFPA 1021: 4.2.2, 4.2.6, 5.2.1

Josh is a very enthusiastic fire fighter when it comes to sharing his knowledge with others, but he often fails to complete routine station assignments. When Josh receives your assignment to put up the station flag and wash the reserve engine, he typically starts talking to the less senior fire fighters on the apparatus floor about the features of the reserve engine or the quirks of the pump panel. This leads to an informal pump operations or hose-handling training session or even a dissection of department policy about response procedures. When walking through the apparatus bay 30 minutes later and passing by the watch office door, you observe the station flag still lying on the watch desk, and the reserve engine remains unwashed.

1. If you do nothing, what will happen?

Station assignments will continue to be unfulfilled. Additionally, you might have a budding mutiny on your hands if the other fire fighters feel that Josh is getting by with less work and you are not addressing the situation.

2. What do you say or do?

Break the scenario into its components: Josh's enthusiasm for training other fire fighters is one issue. Stop—this issue stands on its own. No matter which other issues may coexist, recognize that this one has nothing to do with the others. However, recognize that you can use Josh's enthusiasm to your advantage in the big picture.

The second issue is the fact that Josh is not completing his routine station assignments. Stop—this is another stand-alone issue. When considering these issues independently, it becomes clearer that as the supervising officer, you really have no need to tackle the training issue unless you are using it as an incentive (more on that later). However, the issue of incomplete assignments really must be addressed.

Your first step is to ensure you have clearly identified your expectations to Josh. Consider whether you need to clarify for Josh that the assignments are his responsibility. Are your deadlines clear? Did legitimate issues keep him from completing an assignment? Was he detailed to another station for a period of time? Was he absent due to illness? If these variables are clearly explained and it is simply a matter of incomplete work, pull Josh aside and directly identify your expectations, his performance (or lack thereof), and the consequences if he does not meet your expectations. Further explain any remedial instructions and consequences for his failure to complete assignments after the counseling session. At a minimum, document the discussion for your files and possibly Josh's personnel file, depending upon whether this is a reoccurring issue or other performance shortfalls have been noted.

3. With whom do you address the issue?

You should have a discussion privately with Josh. During this discussion, it would be appropriate to use the other issue provided in the scenario (Josh's enthusiasm for training newer personnel) as encouragement for behavior modification with reward versus punishment. Explain how important it is to set a good example for the new fire fighters not just in the classroom but also in the completion of routine daily tasks. You can acknowledge the positive training influence and noticeable enthusiasm Josh has for training others, but encourage him to improve his own routine performance to enhance his effectiveness and credibility as a trainer.

4. What are two other alternatives you could use?

One alternative would be to identify a new training opportunity to be assigned to someone within your crew, and explain to Josh that you *want* to give him the assignment but you cannot do that until you have seen improvement in the completion of his routine assignments. You must communicate this message with a tone of genuine desire to assign the training assignment to Josh, rather than an implied threat to improve or else! Additionally, this training assignment should be specifically for your crew, not a detail to a different company or division. As a general rule, moving a problem to someone else is not really addressing your responsibility as an officer. Fix the problem or address the issue. Then, when that step is complete, if or when Josh is moved or transferred, you can be confident that you are not simply passing an unresolved problem off to someone else.

A second alternative might be to consider rearranging routine assignments for your entire crew to put the best-fit assignment with the most interested person on your crew. This could be a challenge because if others in your crew are happy with their current assignments, failure of one person to complete his or her share of the workload should not necessitate changes for the rest of the crew. However, if a certain fire fighter is naturally detail oriented and meticulous, you might choose to assign the check-sheet work to him or her and leave the general duties to someone who does not seem to be as concerned with the details. If you can make training-related assignments to Josh, that may be to everyone's advantage.

5. What is your last-resort action?

The last resort may be to not address Josh's enthusiasm for training but instead to focus solely on his poor performance regarding assigned tasks. You may have to spend more time following Josh around and making sure he understands he is "under the microscope." This will create some work on your part as the supervisor, but you will communicate the message that you are serious about everyone doing his or her part. This oversight reassures your crew that you will expect equal performance from everyone, regardless of which additional fire service interests they may have. Send a clear message: "Do your assignments well first, and only then seek additional responsibility."

6. If you reach your last-resort action, what effect will you be dealing with two weeks after taking action?

You will likely have an upset fire fighter who feels you don't trust him, don't like him, and don't appreciate his contributions as a mentor to younger crew members. You will also likely have the rest of your crew smiling and appreciating you for addressing the issue instead of letting it continue unabated. They might even encourage Josh through his behavior modification so he can be relied upon to take on new training duties or tasks. Deep down, most fire fighters just want a fair and equal opportunity for everyone.

5. Instructor Teaches Incorrect Material

▸ NFPA 1021: 4.2.3, 4.2.6, 5.2.1, 5.2.2

Aaron is a senior fire fighter and certified instructor assigned to train others in the department. While sitting through one of his classes, you realize that he is teaching incorrect information. Today he is instructing a multicompany drill in which he emphasizes the importance of salvage and overhaul. He teaches responders to remove all contents of any structure fire down to the stud walls to ensure that the fire is overhauled properly and that fires do not rekindle. He quotes from published training texts about the importance of overhaul and adds that fire investigators need to understand the hard work of response companies. He assures the class that any fire investigation evidence can simply be rebuilt outside the structure for photo or evidence collection for a fraction of the effort fire fighters spend battling the blazes.

1. If you do nothing, what will happen?

If you do nothing, the incorrect information becomes baseline knowledge for the students. This translates into future performance inaccuracies and possible humiliation for them when they are confronted with a challenge, and they perform "correctly" based on what Aaron taught them. Concerns include Aaron's loss of credibility as the instructor, loss of credibility for the training program itself, and ultimately poor performance for the department. If you do nothing, you are also setting Aaron up for failure by allowing him to continue teaching the incorrect information. Finally, if you do nothing despite being aware of the inaccuracy through first-hand observation, you are not performing your responsibility as a supervisory officer for quality control of your department. The consequences of doing nothing are substantial; therefore, you should not remain silent.

2. What do you say or do?

First, do your homework. Double check your facts, and make sure what Aaron said was truly inaccurate. You can do so by conducting research using an objective source—a textbook, a standard operating procedure, even a direct verbal order—but make sure you have a point of reference before you discuss the inaccuracies in the presentation. For example, if Aaron instructs students to overhaul the scene by removing the room's contents down to the stud walls, it would be helpful to draw Aaron's attention to the total published work on this subject, which is likely to include references to preserving fire scene evidence and preventing evidence spoliation, which are also fire department responsibilities. Even if the textbook or internal operating procedure states that fire fighters should remove and extinguish all smoldering materials to prevent rekindle, Aaron must understand the context and the total picture rather than instruct on independent pieces of facts, correct as they may be independently.

Second, think about the long-term solution. Your first objective is to make sure Aaron is aware of the inaccuracy so it does not happen again in the future. You also have an objective responsibility to ensure that all students have accurate baseline information for their development and use in the future. Finally, you have a responsibility as an officer to develop your people. You must correct the inaccuracy while using tact and diplomacy to assist Aaron in becoming a better instructor, rather than seek to humiliate him for making a mistake. Criticizing the instructor is not beneficial to either his growth or his students'.

3. With whom do you address the issue?

After fact checking, have a private conversation with Aaron regarding what you have observed first hand, and cite the factual sources that indicate what he was teaching is inaccurate. Allow him to explain. It could be a case of "garbage in, garbage out"; that is, Aaron might simply be teaching the inaccurate information that he was taught himself. In any event, it is paramount to impress on him the importance of preparing his teaching materials so source information is validated and true for the sake of both the students' learning and Aaron's credibility.

You do not need to have the initial conversation with the students or the peer crew members present. Allow the instructor a graceful exit strategy to make a correction on his own. Direct the instructor to set the record straight for the students. The issue at hand is not highlighting the instructor's mistake; rather, the issue is ensuring students receive the correct information. Ask Aaron to focus on that objective, and then allow him the opportunity to make it happen.

4. What are two other alternatives you could use?

If Aaron is reluctant to make the correction, an alternative might include distributing a training summary sheet to supplement the instruction. Be sure to include the topic or topics that Aaron originally taught incorrectly, and make sure the summary sheet contains the correct information. As an officer, you can send such a training supplement without raising too much suspicion, although it will not take long for the students to discover inconsistencies. Even so, the documentation will indicate the correct material or process, and that message is more likely to pass the test of time.

A second alternative might be to hold a refresher class in the near future in which a different instructor is assigned to teach the subject. If this option is implemented, it will be important to share with the second instructor the situation you are trying to discreetly correct. Make sure he or she draws attention to the accurate information during the class so the correct method is covered—in this case, educating fire fighters about the need to effectively overhaul a fire scene while preserving the original fire scene and contents as much as practically possible. This introduction of contradictory information may create some heated discussions among the students, so be prepared for that outcome. Hold to the facts and the factual sources.

5. What is your last-resort action?

If you have privately discussed the situation with Aaron, allowing him the opportunity to make the situation right with minimal attention, and he has not capitalized on that opportunity, and if the use of refresher training with another instructor is either not available or inappropriate for some reason, then you may have to personally correct Aaron's inaccuracies with his students. Alternatively, you may have to follow up with each of the students in the class that day, having a private or semi-private discussion regarding the inaccuracies and providing the correct information for their use.

6. If you reach your last-resort action, what effect will you be dealing with two weeks after taking action?

Aaron, as the instructor, will most likely resent you for publicly correcting him. The focus here must be on the objectives with which you started: First, you want to inform Aaron of the inaccuracy. That objective was met. Second, you want to provide the correct information for the students. If Aaron chose not to do so, you as the officer have a responsibility to fulfill this objective. Third, you want to develop your instructor. If Aaron did not respond to the opportunity you offered him, development of this instructor requires that he learn the consequences for not correcting his own mistakes—in this case, public correction from a supervisor in an instruction forum. Do not call the group together specifically to make a public correction; rather, make the correction during the next scheduled class. This preserves the integrity of the training program, the department, and you as the officer.

6. Fire Fighter Brings Non–Fire Department Work into the Station

▸ NFPA 1021: 4.2.2, 5.2.1

Chip is a good and effective fire fighter on your engine company, who works in construction as a general contractor on his days off. In the evenings during his shifts at the station, after dinner is finished and the kitchen is clean, Chip tends to spread out job plans and blueprints for the next day's contractor work. Everyone at the station likes Chip, but when he spreads out his documents, he prevents the other crew members from using the table to eat a snack, work on a fire department project, or even study fire training materials. This is not happening just once in a while—it occurs almost every night when Chip is working. You observe that Chip's plans take up almost all of the table, forcing others to go elsewhere or wait until Chip is finished, a few hours later.

1. If you do nothing, what will happen?

If you do nothing, Chip's behavior will continue. The deeper message being sent if you do nothing is that work outside of the fire department is more important than the work of the fire department itself. As an officer, if you do nothing, you are communicating that it is permissible to focus on other things rather than the work of the fire department.

Some might argue, "If the action is occurring after dinner, it's the fire fighter's time anyway. Besides, isn't spending time on outside work more productive than watching television or playing game systems on the station's big screen?" That point suggests that Chip's choice involves the lesser of two "evils." Yes, work is more productive than television, but the question remains: Is either television watching or side-job work productive to the community that is paying for the fire fighter to be there? The issue for you to consider as the company officer is not the specific isolated act but rather Chip's repeated actions—that is, the pattern—and their subtly communicated message regarding priorities. Chip's ongoing appropriation of the common space for his own personal use is also disrespectful of and unfair to other crew members.

2. What do you say or do?

Determine whether any department standards or regulations speak to this issue, and consult those guidance documents. As an officer, you have a responsibility to research, communicate, and enforce departmental policy.

Next, observe what Chip is working on. Don't jump to conclusions, and don't fall prey to rumors or peer pressures exerted by other crew members. Determine if this behavior is an unusual short-term issue or has become a new routine for Chip as he tries to stay ahead of his workload outside the fire department. If the situation is short term (a week or less), then you can seriously consider doing nothing other than monitor the situation. Be ready to jump in and free the table if it is needed for department purposes before Chip's peers have an opportunity to raise the issue.

If you determine this situation is the "new normal" around the station when Chip is working, then a conversation is merited about the need to focus on the community's best interests while at the fire station rather than the best interests of any individual fire fighter's outside employment responsibilities. Down time is an inevitable consequence that comes with the territory when work shifts extend beyond the traditional eight-hours-a-day schedule, but down time is not intended to provide working time for outside employment. Using it for private business is a virtual "double dip" for income. Some fire fighters would argue, "Who will know, and do you really think the public would even care?" All too often, we fool ourselves into thinking the community will not care because we, like Chip, want to get the work done. In fact, many citizens would be outraged! Perhaps some would be indifferent to this behavior, but they will be rare. Most civilian employees in other occupations would be quick to point out they do not have the ability to work on side jobs while they are on the clock for their primary employer.

3. With whom do you address the issue?

Communicate with Chip first. Give him the chance to hear your message and make his own correction without much fanfare. After some time has passed, it may be worth discussing the issue with all of your crew as a teaching point to keep similar situations from occurring in the future. Discuss the importance of using all station time—whether as a career or volunteer fire fighter—to work for the community either directly on responses or indirectly through training, preparing, physical activity, or applying your talents and abilities to find methods to improve the service to the community. Consider using Chip's changed behavior as an example. Positively reinforce Chip for recognizing the importance of focusing on fire department work while at the station. This action accomplishes three things. First, you publicly praise a member of your crew—which is never a bad thing for the crew or you as their officer. Second, you reinforce Chip's behavior to deter him from falling back into his old habits. Third, you reinforce the priority of "fire department business first while at the fire station" to all members of your crew, and you do it in a constructive manner.

4. What are two other alternatives you could use?

One alternative might be to hold a surprise activity for the crew without notice that requires everyone to react. This activity might take the form of an impromptu training evolution on turnout time or driver training. Make the exercise legitimate and helpful to your company's development, but try to have something where you will be out of the station for 30 to 60 minutes or more. Allow Chip to recognize that the fire service is a business where no advance notice is the norm, and to realize that leaving "other work" out on the station table unattended may not be the most responsible action. This alternative will work effectively only if you follow up with a discussion with Chip to point out why he needs to stop the behavior.

A second alternative might include assigning additional responsibilities to Chip, since he apparently has ample free time to work on his outside work. Caution: Be careful not to add work assignments to his workload while you allow fellow crew members to watch television or attend to other personal business. Obviously, working on outside employment responsibilities typically is associated with personal financial gain, but do not let others receive individual personal gains from your assignments either. If the crew has that much free time available on a regular basis, be a proactive fire officer and consider enhancements to the training, physical conditioning, or station/apparatus needs to keep the crew focused on public service while on duty. This applies to both paid or volunteer personnel.

5. What is your last-resort action?

The last-resort action might include applying a more direct leadership style by simply directing Chip not to bring outside work to the fire station. You as the officer have the responsibility and authority to direct the work of your subordinate personnel. Giving such an order would be considered in the best interest of public service and, therefore, would be merited. This option is considered a last resort because although directives are to be followed, if the buy-in or understanding is not there to accompany a directive, you will create a tense atmosphere in which people take sides, potentially leading to conflict. Additionally, people who are forced to act typically perform only to the minimal point required to meet their obligations and will often look for opportunities to relax their performance as much as they can get away with it. You do not want to make the situation worse.

6. If you reach your last-resort action, what effect will you be dealing with two weeks after taking action?

This effect is somewhat determined by the individual fire fighter involved and the culture in your fire department. Some fire fighters would accept the directive for its good intention and promptly comply with the request. Other fire fighters might draw a line in the sand to determine sides and try to assess, by talking with every fire fighter who will listen, who is right and who is wrong by popular vote. Don't be surprised if popular opinion does not favor your decisions as an officer. Be prepared for complaints about the fire department's insufficient pay structure and the necessity for many fire fighters to maintain outside employment to make ends meet. You may even be accused of actively hindering your crew's attempts to get ahead by not allowing them to complete outside work on company time. Remember: This issue is not about Chip's outside employment, his workload there, or his ability to earn income. Rather, the pertinent issue here is about the expectation that community servants will work for the community when they are being paid (paycheck or volunteer incentives/stipends) by the community to serve in that role. It is also about your responsibility as an officer to supervise and enforce this public trust before it is questioned or in need of remediation.

7. Fire Fighter Engages in Non–Fire Department Work During Training and Responses

▸ NFPA 1021: 4.2.3, 4.2.6, 5.2.1

Marcus is a fire fighter on your engine company and also a general contractor for his own construction business, which he runs on his days off. He receives numerous phone calls throughout the day on his cell phone regarding his construction duties and subcontractors. He even receives some calls during training and on emergency responses. When he first joined your company months ago, Marcus would apologize for interrupting a training session, but now he simply excuses himself from the session and steps out of the training room or away from the hands-on activity to talk through the issue on the phone. The calls last for several minutes, and recently Marcus has been observed making outbound follow-up calls after the initial interrupting phone call. In addition, Marcus carries his phone with him on the engine because he says some calls about concrete or coordination are time sensitive. He even took one call on an active alarm response while the company was staged awaiting assignment.

1. If you do nothing, what will happen?

If you do nothing, the situation will only get worse. The fact that Marcus is receiving numerous phone calls means this is not a unique "emergency" in need of a little latitude. Instead, this is a situation in need of your leadership to set the correct course for Marcus and to ensure the integrity of your entire crew, and even your department. If you do not address this situation, it may erode the function of the department because it is a distraction for the rest of the crew. Doing nothing is not an option in this situation.

2. What do you say or do?

This situation allows you as the officer to have some flexibility with what you do. If Marcus is a good fire fighter and a competent performer with a positive attitude other than these frequent phone interruptions, you may opt to handle things discreetly and with a soft-sell approach. Conversely, if Marcus is lacking some, or many, of these basic firefighting skills, a hard-line, direct approach may be merited. Remember, as fire fighters we must focus on the business of the fire department when we are on duty at the fire station. Allowing outside job duties to interfere with fire station business is unacceptable and inappropriate. In this scenario, Marcus is answering phone calls during training sessions and even responses. You need to help Marcus reprioritize the demands on his time. As the supervising officer, you have the authority to instruct Marcus that he is not permitted to take or make phone calls during training sessions or on responses. At other times in the station when personal activities are allowed, you may allow Marcus reasonable time for communication.

Emphasize to Marcus the importance of focusing on preparation and personal readiness regarding fire service responsibilities and his role as a member of the fire service. What Marcus opts to do with his time away from the firehouse is entirely up to him, but when on duty at the fire station, he must remain focused on the business of firefighting.

3. With whom do you address the issue?

Talk directly with Marcus. Some readers might protest: "How can we enforce this directive if Marcus is using his personal cell phone rather than the fire department phone?" But if Marcus was scheduled to be at the fire station (career shift or volunteer station time), you would not allow him to leave whenever he likes just because he owns the car he drove to work, would you? You can set realistic limitations for the personnel whom you supervise *if* those limitations are for the good of the order of the department, your crew, or the individual fire fighter's safety. Additionally, the community expects its public servants to serve the public while on the community's time or dime. There is also an expectation that officers of the department will ensure that this focus is maintained. If Marcus is focusing more on the "other work" phone calls than on his responsibilities as a fire fighter, you need to help him refocus by imposing limitations.

As mentioned earlier, if Marcus is a positive contributor to the team, you might consider a soft approach—for example, informing him privately of the importance of limiting outside work calls or business conversations and *asking* him to comply with the limitations you establish. This approach leaves some decision-making ability with Marcus regarding compliance and ideally allows him to discover the error of his ways without making too big a deal about the behavior. Generally, if the fire fighter is a team contributor in other aspects, the individual will likely correct the situation on his or her own after you have brought the behavior to the fire fighter's attention.

In contrast, if Marcus merely exists as part of the team but does not necessarily actively contribute to the team, or if Marcus is a troublemaker for the team, you might opt to take a stern or direct approach. In this scenario, you would privately communicate the same limitations, but you may need to sternly explain the ramifications of Marcus's lack of focus and the potential consequences for the team and the community relying upon him. You might spell out what is permitted and what is prohibited regarding phone communications while on duty. Inform the fire fighter that he will be monitored, and that immediate improvement is expected. With this alternative, you do not really give the fire fighter an option to embrace the change but rather clearly advise him about your expectations as an authorized supervising officer.

4. What are two other alternatives you could use?

One alternative might be to make a general directive to the entire company. Phone use or abuse issues can start small and develop quickly. You might opt to take a broad-stroke approach by reminding everyone in the company of your expectation that he or she will remain focused on fire department duties. Elaborate on specific limitations such as that no "discussions for profit" are to be conducted while on duty; no phone calls are to be taken or made during training sessions, in meetings, or on responses; and no excessively long phone conversations are to be held that would distract the fire fighters from focusing on their firefighting duties. Specify your expectations for Marcus and the entire crew.

Before you adopt this alternative, you must consider two factors. First, the culprit whose behavior instigates the discussion—in this case, Marcus—should not be a person who will readily take offense about having his mistake called out in front of everyone without having had the matter brought to his attention privately to allow him to self-correct. Second, the tone you take with your entire crew must be one of constructive feedback or coaching for all, rather than a negative or corrective tone. Do not "scold" the entire team if such criticism is warranted for only one person. A refocusing or reminder for all is acceptable if handled constructively.

A second alternative might include an increase in specific types of assignments or tasks for Marcus. This should happen only after you have engaged in a direct conversation with Marcus outlining your expectations and limitations on phone use. At that point, you can reinforce your message by assigning fire department work duties to Marcus to fill his newly available time. If you select this alternative, you must refrain from assigning overwhelming or impossible tasks. The purpose is to help Marcus refocus and to reinforce his good behavior, not to humiliate or discourage him. Remember, he is still part of the team.

5. What is your last-resort action?

The last-resort action might include a direct order from you to Marcus that, effective immediately, he is not allowed to use his cell phone while on duty and that calls on the station phone will be limited in length. Specify to Marcus that all outside business is prohibited on any phones while on duty. This restriction may seem extreme to Marcus, but such an outright ban is a last-resort action. You should issue such a harsh directive only after you have tried a softer approach to gain compliance and Marcus has earned this hard-line position by failing to comply with your previous, less restrictive instructions.

6. If you reach your last-resort action, what effect will you be dealing with two weeks after taking action?

Tension! Some fire fighters may be ready to "take up the cause" by insisting how you have taken away their basic civil rights, such as the freedom of speech. Stay the course. As the officer, you are responsible for keeping your fire fighters focused on their mission as fire fighters. Handling business for profit while at the fire station does not serve the fire fighter, the team, or the community. You should provide an opportunity for self-correction, but if Marcus ignores that opportunity, you have an obligation to mete out the consequences. The firefighting business has enough hazards of its own; do not allow any member of your team to create additional hazards through apathy or distraction.

You will also need to increase your attention to fire fighters' personal behavior to ensure enforcement of your directive. This aspect of the fire officer's job is not pleasant, but it will be necessary to provide accountability. Your leadership will be tested, so be prepared. Eventually, the issue will fade, and everyone's focus will return to the work of the fire station.

8. Fire Fighter Requests Training to Perform Duties Not Applicable to the Department

▸ NFPA 1021: 4.2.3, 5.2.3

Mary submits to you a written training request form, which asks permission to take advanced wildland fire training classes to qualify her for deployment on large western wildfires. This specialty fire training is not applicable to your department's urban/suburban jurisdiction. In the kitchen later that same day, you hear some fellow crew members mention that Mary is preparing to leave this job and find work elsewhere. As you walk back to your office, you consider how to assist Mary appropriately while still acting in the best interest of your department.

1. If you do nothing, what will happen?

If you do nothing, you have failed Mary, who has specifically asked for your assistance. If you do nothing, you not only send Mary the message that you are indifferent to her request, but you also miss the opportunity to discuss Mary's future development in the department.

2. What do you say or do?

This is a good opportunity to meet with Mary to ask her which specific training she wants to pursue and why. Remember—you overheard her fellow fire fighters *speculating* that she is preparing to leave; you did not hear that message directly from Mary. Do not jump to a false conclusion based on idle chatter. After Mary tells you which training she wants to take and why, ask her what kind of assistance she needs from you. You might find that she just needs a letter of recommendation or perhaps some additional study time while at the station. You may be able to assist in these matters. If she requests your approval to pursue training that would take her off duty, or if she asks that the fire department pay for the training, however, you may have a different scenario.

Some people might recommend that you approve the training and send it up the chain to see what happens. This approach keeps you from having to be the bad guy or confront the issue. In truth, the company level is precisely where decisions regarding training should be made. Training is good. Although any training will make a person a more valuable fire fighter than before the training, not all training is relevant or appropriate to be supported with time or money in every department. The company officer must assess whether the training is relevant to the service of his or her particular community.

If the biggest wildfire potential in your jurisdiction is a 30-acre park, wildfire command and control training is probably not appropriate for your department to support; however, the presence of a substantial suburban-forest interface in your jurisdiction might be a good reason to support the training. It is important that you discuss with Mary why she wants to attend the training. You may discover she wants to prepare for future promotion within your department and saw this training opportunity as a good command and control experience. You may be able to support her desire for increased training in that area but in a class more appropriate to the jurisdictional hazards, such as structural firefighting or target hazard-based command and control classes.

If Mary is interested in showing initiative or she wants a new challenge due to burnout or boredom in her current positional responsibilities, you also have the opportunity to support her and provide her with new challenges to keep her motivation high.

3. With whom do you address the issue?

Speak with Mary directly. It may be appropriate to ensure some privacy during this conversation, but it would not necessarily need to be confidential. This type of discussion could occur during some down time during the day—maybe a front-bumper conversation in the bay or a cup of coffee in the kitchen after meal clean-up is complete would be a good opportunity to talk. If you keep the conversation casual, you may avoid putting Mary on the defensive. You want to be open and cordial but also able to give a definitive "no" if warranted.

You might also have a private conversation with your supervising officer after you have obtained all the details from Mary. Keep your supervisor informed of the aspirations and challenges of your crew, as well as what you are doing to develop your crew. Your officer will appreciate knowing you are willing to say no to company requests that are not congruent with the fire department's mission.

4. What are two other alternatives you could use?

One alternative might be to ask Mary to research and provide you with all the costs associated with her request, including registration, travel, housing, meals, station coverage while she is away, and so on. When she provides that information, you can assess how those figures match up with available training budget funds. Sometimes it becomes clear that even a good opportunity for one fire fighter must go unrealized if it would require 50, 60, or even 70 percent of the entire training budget intended for all. You will need to research which funds are available and how widely they are to be applied so that you can discuss what is available or not available at the follow-up meeting with Mary.

A second alternative might be to ask Mary to write a training request outlining the costs mentioned earlier, and then add a narrative component explaining specifically why this training would be a good investment for the community to make in her. Fire department training is funded generally by local money—taxes, fund drives, grants, and so on—so ask her to explain how this training request will benefit the community you serve. Be prepared to discuss Mary's submission with her. If she really wants to attend this training, she will likely find some creative ways to justify it. The activity of creating the written narrative is designed to make her really think about the fit with the taxpayers. In many cases, a poor fit becomes painfully clear even to the requestor, and such a realization might prompt Mary to withdraw the request.

5. What is your last-resort action?

Just say no. It is never popular or pleasant to say no to training of any type. In fact, you should not make a habit of arbitrarily saying no to training, especially specialized or advanced training. However, if you have doubts regarding how the training will benefit taxpayers or how the training might be applied toward personal gain outside the job, the safe stance is to say no, for now. Rest assured that other training opportunities will arise in the future. Use the time provided by this "no" response to gather more information about the aspirations and intentions of your crew so that only appropriate requests come to you. As a fire officer, you should help develop your crew and their opportunities to the best of your ability with the resources you have available.

6. If you reach your last-resort action, what effect will you be dealing with two weeks after taking action?

Depending on the personalities of those involved and the effectiveness of your communication with Mary, the effect two weeks later may be positive for all involved. Your crew—and specifically Mary—will know you care about helping her get training, but they will also know you are a responsible leader who seeks to do the right thing for all involved.

If your communication is poor, or if your crew or station is populated with conspiracy theorists (yes, they exist everywhere!), then some hostility may linger in the shadows for a while. There will be rumblings that you are trying to keep Mary down, or insistence that you just want to protect your job, or even tall tales that you refused the training request because you do not like Mary and you wanted to show her who the boss is. The key here is to ensure effective communication with Mary, which sets the stage for positive messages to spread, despite the "no" answer to Mary's request.

Interpersonal Dynamics Issues

9. Inappropriate Comments to a Peer Fire Fighter

▸ NFPA 1021: 4.2.4, 4.2.5, 5.2.1

Scott is a senior fire fighter on your ladder company. Scott is poking fun at a fellow crew member, Roger, the new recruit assigned to your ladder company earlier this year. Scott starts joking with Roger about his forcible entry skills but then changes topics to ask Roger about his personal life. After hearing that Roger is not married, Scott sees an opportunity to make Roger squirm a little and begins to question him about the women he has dated. Scott starts commenting on the women's names, hair colors, and interests, as well as the restaurants where Roger has taken his dates. Roger mentioned recently that he just started seeing someone new and is taking her out to dinner at a local restaurant the following night. Scott sarcastically tells Roger that he needs to improve his choice of restaurant if he wants to "land a wife." As the officer, you have overheard bits and pieces of conversations before, and now you are hearing more and more of the story through crew members' conversations in your presence about Scott's comments to Roger.

1. If you do nothing, what will happen?

Possibly, nothing will happen. The situation may potentially remain a case of friendly joking among your crew. However, it is also possible that Roger is bothered by the teasing, but as the "new guy," he is hesitant to complain. The key here is recognizing that as an officer, you have a responsibility to pay attention to the situation and determine whether it is now a problem or could rapidly become a problem and to act quickly to head it off. If you are not sure, it is better to err on the side of preventing a possible problem from developing rather than waiting to see.

2. What do you say or do?

Sometimes an officer injecting a comment into the discussion is all it takes—no discipline, no fanfare, just a redirection of the conversation from you as the company officer explaining the subject is off limits is all it takes. You might say something like "Okay, Roger's personal life is his business, not yours, Scott, and we will not be talking about it here. Our focus is on emergency services, not dating services." This is firm enough to convey your expectations while keeping your correction low key.

3. With whom do you address the issue?

In this situation, you may decide it is best address the issue in front of everyone. This approach keeps the situation from appearing as the counseling session that a one-on-one conversation would create. It also sends the message to everyone within earshot that you have marked that conversation as off limits.

4. What are two other alternatives you could use?

First, you could pull Scott aside privately and tell him to stop. This option will appear like a counseling session, but you can keep the tone conversational to reduce that perception. Second, you may talk with Roger and inform him to avoid sharing such personal information so as not to give Scott "ammunition." If Roger did not personally share information about his dating life, and it was the other crew members who discovered it on their own or are merely speculating, then the direct approach with Scott may be most appropriate.

5. What is your last-resort action?

If you tell Scott to stop discussing Roger's personal life, and he continues to do so, Scott is guilty of insubordination. You have to deal with the insubordination appropriately, and those steps may be far more severe than the desired course of action from the initial comments.

6. If you reach your last-resort action, what effect will you be dealing with two weeks after taking action?

This action will likely create resentment among the crew and toward you as the officer. Being a company officer is not easy. You will now have to work to repair hurt feelings, and this could negatively impact job performance. Given this possible outcome, some officers might argue that dealing with such passive comments simply creates more work for the officer and damages the crew relations, so it is best to just let the behavior go. The final question to ponder, though, is this: If addressing inappropriate comments strains the crew relations to that degree, did you really have such a tight crew in the first place?

Dating issues, marriage issues, and other personal relationships can start your company's trek down the road to many complex legal issues. Laws are in place that protect people based on sexual orientation, political preferences, race, ethnicity, and many other factors. What seems like innocent joking around to many fire fighters could actually be the beginning of a civil litigation situation where someone feels discriminated against or feels the workplace has become hostile because of the individual's personal choices. Many fire personnel might argue, "That's just the usual station antics. If they aren't picking on you, then they don't like or accept you." In reality, the fire service culture doesn't relieve the department from the responsibility of ensuring such harassment does not happen. This understanding is a critical part of your role as an officer: You have the responsibility to detect and intervene to stop such behavior. You are the one who hears it, sees it, or becomes aware of it first. If you are not aware of it, then you need to be more observant with your crew. Denial of the responsibility will not get you or the department off the hook.

10. Fire Fighter with Skill Deficiency

▸ NFPA 1021: 4.2.3, 4.2.6, 5.2.1

You are a company officer in a combination department. Keeping career and volunteer members aligned is a constant challenge. A career member, Jill, comes to you to report that one of the volunteer members, Dan, is not proficient in basic firefighting skills, such as breathing apparatus emergency procedures, hose stream selection and safe use, and even fire behavior recognition—all of which are necessary skills to perform successfully as a fire fighter. Jill tells you that she does not trust Dan to work on emergency calls with her.

1. If you do nothing, what will happen?

A lack of trust is a serious matter. If you do nothing, you allow the culture of distrust to take root and grow. It can create factions and divisiveness that can destroy a combination department. If you do not get to the root of the issue, the two individuals involved will constantly be looking for problems with each other instead of focusing on the team. You cannot allow this to happen. You, as the company officer, must deal with this situation.

2. What do you say or do?

First, you must investigate the matter for yourself to see where the facts begin and the fiction ends. The career member, Jill, may be absolutely correct. Alternatively, Jill may be trying to find fault with Dan, a volunteer who is not present as much as the career member. Sometimes a reported lack of skills is a general statement implying lack of dedication, training, or commitment. In some organizations, it could be a push to replace volunteers with more career staff. If a skill deficiency truly exists with Dan, you owe it to Jill and the rest of the department to address the issue quickly and appropriately. If you discover a personal agenda or a "witch hunt," you owe it to Dan and the entire department to address the problem promptly with Jill.

In this case, we will assume the skill deficiency does exist. You can talk with Dan and establish a coaching plan to make sure his skills are learned, practiced, and—most importantly—assessed to ensure he is proficient. The key is to make this a coaching experience, not a test or hostile environment. Some officers might protest that they do not have time to coddle "slackers." Lives depend on fire fighters having proficient skills, but no one is born with proficient firefighting skills. Everyone had to learn these skills for the first time. Some people learn faster than others, and more senior members of the fire service tend to forget what it takes to master the basic skills and reach the necessary proficiency levels for the first time. The officer must provide Dan with every opportunity for successful learning.

If Dan cannot demonstrate skill proficiency, even after coaching and additional training, he should be disallowed from participating in a capacity requiring those skills. If the skill problem is more fiction than fact, however, you must deal swiftly and severely with Jill. There is no room for fiction or fault finding among fellow fire fighters when it comes to skills. No one has time to repeatedly sort through false reports, so you must send the message that fiction regarding fellow fire fighters is not tolerated.

3. With whom do you address the issue?

You should deal directly with the individuals involved, Jill and Dan. You may need to involve a training officer or assign a trusted fellow fire fighter to serve as a mentor for the volunteer if you have too many other responsibilities to dedicate sufficient time to developing Dan's skills. In any case, talk to Dan and Jill frequently and with a coaching perspective. Both need to know you are staying on top of the situation and you care about both perspectives for the greater good of the department.

4. What are two other alternatives you could use?

If you have given Dan ample time and opportunities to improve his skills, and he has not applied himself to those opportunities, you could reprimand him for not being adequately prepared in required skills. Be aware that this action will send a shockwave through the volunteer force, so be ready for those who may be uninformed regarding the facts of the situation to react negatively and rally around Dan and all the volunteers.

Another option would be to use the situation as a catalyst for developing and quickly implementing a basic skills proficiency assessment for all personnel—career and volunteer alike. This provides an equal playing field and will reveal any skills inadequacies in the department. As the officer, you may want to start with just your crew and the volunteers with whom you interact. You can suggest through the chain of command that such a concept be considered for the entire department.

5. What is your last-resort action?

A last-resort action might be having a direct discussion with Dan. Inform him that a concern was brought to you regarding his skill level, and discuss the implications for the other fire fighters and the department. Allow Dan to explain his position. You may have to recommend suspension of riding privileges until Dan can meet the skill requirements.

6. If you reach your last-resort action, what effect will you be dealing with two weeks after taking action?

The last-resort action will likely put Dan on the defensive. If Dan feels threatened, he may react in any number of ways. In two weeks, you will likely have strained career-volunteer relations. Everyone will examine everyone else's skills under a microscope, and it is possible they will all be expecting you, the company officer, to address any newfound deficiencies. Try to anticipate this reaction and use it as an opportunity to strengthen your team communications. If you reach your last-resort action, you may have a major human relations issue on your hands rather than simply a skills issue.

11. Career Fire Fighter Questions Volunteer Fire Fighters' Skills

▸ NFPA 1021: 4.2.3, 4.2.4, 4.2.5, 5.2.1

While sitting in your office late one night after a training session, you are approached by Frank, a longtime volunteer member. Frank tells you that you have to do something about Tony, a recently hired career member. Frank explains that Tony "struts" through the bay with the other volunteers in attendance for training and is often heard telling the volunteers that if they work hard, they, too, can someday be hired and learn how to be a "real" fire fighter and not just a "play fire fighter" when a call comes in. Frank perceives that Tony is arrogant and condescending and is concerned that Tony's attitude is negatively affecting the company.

1. If you do nothing, what will happen?

It will get worse. When dealing with perceptions and feelings, you will find that "what gets left alone grows in the dark." There may be no basis for the statement, but as the officer, you need to do something.

2. What do you say or do?

First, be appreciative that Frank brought the concern to you to handle. Thank him for coming to you rather than stewing over it. Affirm Frank's behavior by emphasizing that his actions demonstrate his trust and respect within the organization. Next, listen intently and ask questions to get all the facts from Frank's perspective. Let Frank know you will look into the matter and get back with him. Then observe Tony's behavior objectively for yourself. If necessary, based on your findings, pull Tony aside and talk with him about the issue. It is best to not disclose the source of the complaint; rather, include as many examples from your personal observations as possible. It will be difficult for Tony to challenge you as the company officer when you can present direct observations. You will need to coach Tony through the concept of human relations and knowing how to deal with people, including respect for others simply for being members of the same team. Instruct Tony that respecting others is not a request but rather a requirement of the job.

3. With whom do you address the issue?

Talk first with Frank, the affected volunteer. Then talk with Tony, the affected career member.

4. What are two other alternatives you could use?

You could facilitate a meeting between the two members to work through their perceptions and actions. Be aware that this could be a very tense and uncomfortable meeting. If you choose this option, the officer must be very comfortable with conflict management and a very savvy facilitator. Despite the potential risks, this option may be the most direct method of addressing the situation.

A second alternative might be to put the two members, Frank and Tony, together for some drills or projects. Sometimes members cannot stand each other while interacting only briefly in the department but then discover when they are forced to work together that the other person really is not as bad as imagined; as a result, they work things out on their own. This option will require you to demonstrate a high degree of involvement by closely monitoring the situation. Matters can go from bad to worse quickly, so be ready to separate the two fire fighters and use another alternative if the situation deteriorates. Don't expect a complete transformation into a close relationship, but time spent working together on a common goal or project may be helpful.

5. What is your last-resort action?

Sit the two members down separately, and advise them that they are required to work out their differences and that they must focus on the mission of the fire department if they want to continue to be there. This threat is extreme, but it sends the message that the company cannot tolerate their inability to work together.

6. If you reach your last-resort action, what effect will you be dealing with two weeks after taking action?

You may have to suspend or dismiss one or both of the members if you reach this point. When someone places a higher priority on his or her attitude than working together as part of a team to serve others in emergencies, that individual reveals his or her true personality. This self-serving attitude is not what your company, your department, or your community needs. As a company officer, you may not have the authority to suspend or terminate an employee, but you will likely be the person who takes the first steps by filing the paperwork to start a disciplinary process.

12. Conflict Between Two Fire Fighters on Different Shifts

▸ NFPA 1021: 4.2.2, 4.2.4, 4.2.5, 5.2.1

Two members of your crew—John, an apparatus operator for your rescue company, and Sarah, a fire fighter on a different shift but in the same station—are in personal conflict. John and Sarah have an oil-and-water relationship: Everything Sarah says seems to grate on John's nerves, and vice versa. Although John's and Sarah's interactions are mostly limited to those occasions when they pass each other during the shift change every third morning or when a rare overtime shift or shift trade occurs, these moments are tense. Even the shift trades are infrequent because neither John nor Sarah really wants to spend a shift together if it can be prevented. When they do speak to each other, the tone is pointed and sarcastic, and the exchanges are contentious. Light conversation about food or cars ends in discord. What one likes, the other seems to hate. As the officer, you have observed this negativity between your two crew members first hand, but other rescue company crew members have also come to you to express their concerns. Although John and Sarah typically interact only during a brief shift change procedure, you are keenly aware that the ensuing tension sets an unwanted tone for your crew for the day.

1. If you do nothing, what will happen?

If you do nothing, matters may get worse, stay the same, or even improve depending on all the outside factors. Doing nothing is a dangerous option for you as the company officer. Abdicating responsibility to chance, especially when it comes to supervision of your crew, is dangerous because leadership is not a matter of chance. As the officer, you are duty bound to lead and supervise your crew members to ensure an effective workplace. The danger comes from substituting chance encounters and hopeful thoughts that a situation will improve on its own for your responsibility to influence an effective outcome. When as an officer you choose to do nothing, you abdicate your responsibility—and that decision carries negative consequences not only for your crew but also for your effectiveness or perceived lack of effectiveness as a fire officer.

The challenge here is recognizing the conflict has come to rest on your shift, in your station, with John, your crew member. Although you may not have all of the information, or any idea where any responsibility for the discord can be placed, it has become an issue for John. Moreover, given that this conflict involves at least two shifts, your credibility as an officer is now on the line. When conflict exists at shift change, everyone in that station will be aware of the problem. If you do nothing and the conflict either remains the same or even worsens, you and the fire officer on the other shift could be perceived as complacent or indifferent. Who is the stronger shift officer? The fire fighters may be making that decision for themselves based on what they see—or do not see. You must engage.

2. What do you say or do?

One of the most challenging issues in this case is that the conflict may or may not be work related. Dating relationships, financial deals, or hunting privileges may be to blame or disagreements over constantly dirty apparatus, equipment readiness, or trash left for the oncoming shift. At this point, you have no idea where the problem began.

First, find out what lies at the root of the conflict. Call John in privately to talk. Start with observations you have made personally or specific statements you may have received from your crew indicating that something seems to be amiss. Allow John to give you the facts; if necessary, you may need to prompt him by mentioning it seems to specifically involve Sarah. Gauge where things go from there, and be prepared to facilitate the discussion.

3. With whom do you address the issue?

The person to approach first is your fire fighter, John. If there is a conflict involving John and you approach anyone else first, you lose some credibility with John. If details still seem elusive or unclear after talking with John, the next step might be to ask others who have direct knowledge or responsibility, such as your peer shift officer, about the situation. Be careful, though. The people whom you speak with and your timing can have dramatic effects on the situation—either positive or negative—so proceed with caution.

4. What are two other alternatives you could use?

You could transfer John to a different station to eliminate all interactions between John and Sarah at shift change. This action may solve the problem temporarily, creating the space needed to allow things to work out over time. If John is a valuable asset to your crew, however, such a transfer would create a void in your crew.

Another alternative might be to discuss the situation with your peer company officer and work out a short-term swap to detail both these fire fighters to the same station, on the same shift, with the company officer clearly explaining that the fire service is a team, and personal conflicts must be set aside to ensure the effectiveness of the team. It is amazing how many times conflicts originate from simple misunderstandings. Through prolonged, involuntary interactions, people sometimes discover that perceptions are not always reality, and once strongly held misconceptions or misunderstandings dissolve as time reveals the truth. John and Sarah may not become friends, but at least they may become effective teammates.

The least disruptive alternative for all parties involved would be the direct discussion approach. This is especially true if the conflict has arisen over something outside of the work relationship. A discussion with John encouraging him to refocus on the commonalities of the working relationship may make the situation at least bearable while John and Sarah are present at the station. As an officer, you are not responsible for solving everyone's problems, but you are responsible for making sure John and your other fire fighters do not let personal problems stand in the way of their carrying out their assigned responsibilities as effectively as possible.

Be cautiously aware that reactions to your actions may be varied and wide ranging. They may range from "Finally, my officer is going to take care of this" to "Buzz off—this isn't any of your business and doesn't involve the fire department!" The important consideration for the company officer is to remain objective. Inform your fire fighter that you are getting involved in the situation because the consequences of the conflict are becoming a workplace issue. Behavior, workplace climate, shift relations, or any combination of these factors that leads to observable behavior requires the company officer to at least investigate what is happening with the personnel. You may need to reassure John that your involvement is not intended to judge who is right or wrong. You may not be able to solve the problem that led to the conflict, but you have a duty to keep your fire fighters focused on their work tasks.

5. What is your last-resort action?

The last resort may be to engage employee assistance programs through your department's chain of command. If an issue, despite efforts to talk it through or create space, continues to affect the workplace for other personnel and the individual fire fighter's performance at work, then it may be beyond the problem-solving abilities of an officer of the fire department. An employer referral to some form of employee counseling may be required. This is truly a last resort. Nevertheless, just as some big fires have created some big parking lots within the jurisdiction of some excellent fire departments, fire officers must face the fact that they cannot solve every issue, every time, on their own.

6. If you reach your last-resort action, what effect will you be dealing with two weeks after taking action?

Turmoil! If you are unable to handle things at the company or station level, the issue will become the talk of the department. Many times, the individuals involved will talk about "their side" with anyone who will listen. The tricky part about reaching the last-resort action is that you, as the officer, are bound by confidentiality restrictions. Speculation and "fishing" for information will likely become significant distractions for you if the issue becomes the topic of gossip. As the fire officer, you must do your best to refocus your crew and others with whom you come in contact through normal work assignments and ensure that each fire fighter does his or her part to prepare, respond, and recover from the duties that have been assigned to that member. This is not easy. When you cannot talk about what happened and why, try to focus on what you can control—assignments, proficiency, training, and quality. You may find it necessary to temporarily adjust your crew's schedule to keep them too busy to gossip.

13. Conflict Between Two Fire Fighters in the Same Company

▸ NFPA 1021: 4.2.4, 4.2.5, 5.2.1

Stan is a member of your rescue company and has been an active fire fighter and rescue technician for more than five years. Stan has a personal conflict with Jake, another member of your rescue company. Jake is the apparatus operator for the rescue company and has been under your supervision for 11 months. Stan and Jake simply do not get along. When not in your presence, they speak to each other with a nasty and sarcastic tone. This negativity especially tends to surface whenever they must work together to handle nonemergency business such as cooking, training, or station chores. They can perform assigned tasks effectively but offer no voluntary assistance to each other throughout the shift. On a personal level, they disagree on politics, religion, and many other topics that come up in casual everyday conversation with the other rescue company members. You are aware of the sarcastic comments and tense work atmosphere every shift day, all day long. The personal conflict is affecting the entire rescue company by creating a biting and negative atmosphere during the company's workday. Positive events and encouraging feedback from outside the company are often overshadowed by the constant conflict.

1. If you do nothing, what will happen?

If the conflict is between members of your crew and you do nothing, things will definitely get worse! When those involved in the conflict have no means of escape from it when it arises, the situation will intensify. This conflict will also become your problem because in this scenario, as the officer, you have supervisory responsibility for both members, Jake and Stan, and the conflict presents itself while you are present. To use a size-up analogy, to do nothing would be like arriving at a working car fire, and because the owner says he needs a new car anyway, you direct the engine to drive off. A car on fire, just like station personnel conflict, requires appropriate leadership attention.

2. What do you say or do?

Though doing nothing would create dramatically different results from the previous scenario, what you should say and do is very similar. The conflict may or may not be work related. Dating relationships or financial deals are just as likely to be to blame as disagreement over how to pull the attack line, stow the equipment, or conduct the house chores. Right now, though, you simply do not know where the conflict originated.

First, find out what is at the root of the conflict. Call Stan in privately to talk. Start with your personal observations or statements you may have received from your crew indicating that something seems to be amiss. Allow Stan to give you the facts; if necessary, you may need to prompt him by mentioning that your observations seem to indicate the conflict might involve Jake. See where things go from there. Do the same thing for the other fire fighter, Jake. Do not judge and do not take sides; simply evaluate the information provided from both sides to gain understanding about what is happening.

3. With whom do you address the issue?

Talk to Stan and Jake individually first. Discussing the situation with the rest of your crew is not helpful unless the conversations with the two involved individuals left unanswered questions about which other crew members may be able to enlighten you. If it is unlikely they will have any input—for example, if the conflict is over loaned money for a home repair—gathering input from other crew members will not likely produce any insight for you as the officer and may stir up more controversy.

4. What are two other alternatives you could use?

Help Stan and Jake refocus on their work tasks, and clarify your expectations as their officer that they will put the conflict away while they are at work. This directive may successfully solve the conflict; if it does not, you have solid ground as the officer to reprimand them for disobeying your legitimate orders. If a reprimand becomes necessary, the fire fighters need to understand that this action is not about you flexing your authority but rather represents an attempt to keep your fire fighters' behavior focused appropriately. Conflict is a distraction from the mission.

A second alternative might be to objectively judge the situation and award "right" and "wrong" positions in the conflict. This is really an option only if the conflict stems from workplace issues. Listen objectively to both perspectives, gather crew feedback if needed, and then as the officer make the decision on which way to go. One of the parties may like you and your decision, and the other may dislike (or even hate) you and your decision. Remember—this is not about being liked; your decision is about charting the course for the future and leaving the conflict in the past. As the officer, you have the ability when dealing with company-level workplace issues to say, "I have considered the input and the options, and I have decided this is what we are going to do." Like it, hate it, or leave it—the fire fighters must understand it is the officer's responsibility to make decisions to keep the conflict from distracting the team from their duty of serving the public as fire fighters and as part of an effective team.

Generally speaking, the best alternative is to clarify workplace expectations and privately advise Stan and Jake of your expectations for both of them. Regardless of whether the root cause of the conflict is work related, an officer is on safe ground when clarifying job roles for his or her crew and allowing the members the chance to at least put the conflict in check. Failure of Jake or Stan, or both, to quell the conflict allows you as the officer to deal with insubordinate activity or failure to follow legitimate orders, at which point the conflict issue becomes secondary to this new transgression. You need to ensure that the entire team, your crew, is ready to respond effectively at all times.

5. What is your last-resort action?

Separating Stan and Jake, issuing sanctions for failure to follow orders when specifically directed to focus on work performance rather than the conflict, or even sitting the two involved members down at the table with you and trying to directly talk things out may all be appropriate last-resort actions, depending on the issue. The last-resort action will ultimately be decided by the members involved and whether they can put their work first.

6. If you reach your last-resort action, what effect will you be dealing with two weeks after taking action?

If all goes well, there will be no more conflict. If the parties are separated by transfer, suspension, termination, or just an agreement to keep space between the two members when they are at work, the result may be a conflict-free workplace. Regardless of how you deal with the conflict, there will be scuttlebutt among the crew, but if the focus remains on work, the gossip mill will eventually move on to something else.

14. Fire Fighter Disregards Chain of Command

▸ NFPA 1021: 4.2.2, 4.2.5, 4.4.4, 5.2.1

Bill is a two-year veteran fire fighter of the department and a recent transfer to your engine company. Whenever he has operational or administrative questions, Bill routinely bypasses you as the company officer and solicits answers from the chief officers instead. It is not unusual to walk by the chief's office or the assistant chiefs' offices during the course of any day and find Bill standing in the doorway inquiring about why the department purchases apparatus soap from a particular vendor or which content might be on an upcoming promotion process so he can be well prepared. Bill never brings these types of questions to you, his immediate superior. Sometimes Bill even reports the questions' answers back to you as if to keep you informed. You are frustrated with Bill's ignoring the chain of command.

1. If you do nothing, what will happen?

Doing nothing in this scenario would be comparable to arriving on scene at a single-story residential structure with fire showing from the bedroom window and deciding to park and watch! If Bill is not communicating with you but rather takes matters directly to your officers, his behavior creates a supervision issue that reflects poorly on you and on the entire organization. Doing nothing could contribute to your exit from a leadership position. Failure to act will have some serious negative implications for the organization and for you as the leader.

2. What do you say or do?

This situation calls for dialogue both up and down the chain of command. You must establish clear expectations regarding channels of communication for your subordinate personnel. Many scenarios reviewed previously have revealed how effective a private conversation between officer and subordinate can be. This scenario is slightly different because it involves a breach of a very basic function of organizational dynamics in any fire department—the chain of command.

Most people who are involved in emergency services where rank exists understand the hierarchy of the rank system. This rank system does not indicate a subjective value of a person; rather, it indicates order and process within the department, in an organization whose members literally face life and death consequences. The chain of command must be enforced because of the value of the system rather than to satisfy the ego of any individual participant.

Because of this basic understanding and the seriousness of the need to get it right, you should document the episodes where Bill has bypassed the chain of command. This documentation can take the form of an interdepartmental memo; it does not need to be disciplinary in nature, unless you are sure that Bill had some malicious intent. The memo should clarify the specific details, including times, places, and events that compromised the chain of command's effectiveness. It also allows you, as the supervising officer, to clearly state to Bill that it is your responsibility as his direct supervisor to have the first opportunity to address the issues. In addition, the memo enables you to emphasize that issues you handle effectively are issues that do not require the time and attention of higher ranking officers of the department, which in turn allows them to address their responsibilities regarding the organization's greater needs.

3. With whom do you address the issue?

After drafting the memo, bring Bill in for a private discussion of the memo and the situation. It will be important to point out that the memo is a result of repeated occurrences and not simply a reaction to one event.

Following the meeting with Bill, you should notify the supervisors within your chain of command that this memo has been distributed and respectfully request that in the future you would appreciate superior officers referring Bill back to you as his company officer as the initial step for any issues Bill may have. If the chief officer(s) who have been receiving this communication from Bill have previously made a request of him, then relaying that request again through you, Bill's supervising officer, enables you to keep abreast of any assignment or task as well as assist in keeping the workload distributed fairly among the company. This practice keeps you informed but does not disrespect your superior officers.

4. What are two other alternatives you could use?

You can simply tell Bill that he should stop bypassing the chain of command. This alternative is less formal and less threatening but will likely be less effective as well. The scenario here states that the member "keeps bypassing you." This implies that a pattern has become established. A verbal mention of the request may end up backfiring and become a source of fire station jokes rather than addressing the situation. That said, this option is still an alternative for your consideration.

A second alternative might be to request a meeting with the chief officer whom Bill keeps consulting for questions. In this meeting, you can explain how you are being denied the opportunity to fulfill your responsibility as company officer as effectively as possible because your subordinate keeps cutting you out of the information loop. In other words, this breach in the chain of command eliminates any opportunity for you to address the issue before it reaches the chief officer's level. Explain that for you to fulfill your responsibilities, you need the chief officer's help by having him or her kindly refer Bill back to you as Bill's company officer and the first person to address his questions and concerns. If the issue merits the chief officer's assistance, then you should communicate that need through the chain of command with your thoughts or questions accompanying the member's.

5. What is your last-resort action?

Your last-resort action may be to pull your crew together and clarify the chain of command for routine and emergency issues. You can do this in a way that avoids calling out Bill as the offending member in front of everyone, yet still communicates your expectations for all in your command. For those who have been following the proper process, this will not be an issue. For those who have not, it puts them on notice about your expectations. For Bill, it leaves no room for the "I didn't know that's what you wanted" excuse, because now everyone knows the appropriate process.

6. If you reach your last-resort action, what effect will you be dealing with two weeks after taking action?

Some members will wonder what motivated your speech. There will probably be some speculation about what the infraction was or who committed it. Ultimately, this action should have a positive result in stopping the unwanted communications.

15. Fire Fighter Fraternizes with City Manager

▸ NFPA 1021: 4.2.5, 4.4.4

Your department has a chain of command that all members understand and are expected to follow. You learn through passing comments of other fire fighters around the station that Kevin, a member of your ladder company, has become increasingly friendly with the city manager. Kevin did not previously know the city manager, but since Kevin was elected as the fire fighters' labor representative, the city manager has extended various dinner invitations to Kevin and given him tickets to sporting events. In some instances, the only city personnel in attendance were the city manager and Kevin. The city manager's family and Kevin's family have also attended multiple family picnics together. While casually chatting in the station kitchen, Kevin mentions that fire department business sometimes enters his off-duty conversations with the city manager.

1. If you do nothing, what will happen?

If you do nothing, the activity will likely continue. Alternatively, acting upon any type of intervention may yield the same result. This scenario can be a challenge because this is truly an outside-of-work issue. In other words, as the supervising officer, you have no real authority to act. Nevertheless, you must consider what your response will be if your fire chief or one of your supervising officers approaches you and asks if you were aware that your fire fighter is attending social interactions regularly with the city manager. You must acknowledge that you were aware of this activity. Don't miss the point in this situation: It may not be appropriate to say anything to Kevin, but it would be appropriate to mention the situation to your supervisors as a courtesy.

2. What do you say or do?

This scenario is challenging because Kevin is free to make his choices about where and with whom he spends his free time. You should probably at least consider having a casual conversation with Kevin. Mention to Kevin that he should be cautious about talking "shop" with his friend, because it has the potential to put him in a difficult position with his friend, the department, and his peers. The conversation should be brief and focus only on being cautious about what he discusses with the city manager in social settings. You should emphasize that you are not asking which conversations or activities take place, because you admittedly have no need or responsibility to discuss Kevin's personal life. However, you may want to pose this scenario to Kevin: Suppose he were at a social function and the city manager casually asked, "So, how are things down at the station?" or "Hypothetically, if cuts had to be made in the fire department, where would you look first?" How would Kevin respond? How does he think the fire chief would feel about that discussion? How does he think the fire fighters or the local union would feel about that discussion? How would he feel about choosing to respectfully decline to answer the city manager's question? Reassure Kevin that the purpose of your discussion with him is not to dictate his friends or activities but solely to raise his awareness of how this relationship with the city manager could potentially take him to places where those scenarios must be considered.

3. With whom do you address the issue?

Obviously, this discussion would need to be private, between just you and Kevin. You might strategically choose a casual meeting location within the station for this purpose. If you have an office, it might be best not to use it for this discussion. Try to find a private location in a more familiar setting, such as on the bench or chairs on the back porch or in a workout room, or even in the day room, as long as you are not likely to be interrupted or overheard by other personnel in the station. This is not an official supervisor-to-subordinate meeting but rather a friendly gesture to raise awareness to a potential blind spot for your fire fighter. As a caring and effective officer for your crew members, you want to help prevent problems before they arise for the crew or the individual as it relates to the department.

4. What are two other alternatives you could use?

One alternative might be to let the situation unfold on its own and not say anything to Kevin but rather have a brief informal chat with your supervisor. Inform him or her that you are aware of the increased interaction between the city manager and Kevin. Furthermore, state that you have recognized the fire department has no authority or responsibility to determine with whom any fire personnel can and cannot associate outside of work. Inform your supervisor that you have determined the best course of action is to at least inform him or her of the situation; then, if the rumor mill should start up, or if questions are asked by officers higher up the chain of command, your supervisor can explain the facts as they are known. It important to recognize that your supervisor could just as easily be negatively impacted if the fire fighter's relationship with the city manager should deteriorate, even if through no fault of the fire fighter.

A second alternative might be to request, through the chain of command, a brief audience with the fire chief. Inform the chain of command officers that the issue is not a work issue but rather a personal situation that you would like to discuss privately. Should the meeting with the fire chief be granted, you could explain the situation just as you would have to your immediate supervisor. The key to this alternative is to discreetly raise the chief's awareness of the relationship because in most jurisdictions a city manager is the direct supervisor of the fire chief. As mentioned previously, the fire department, including the fire chief, does not possess authority over personal relationships. However, through this meeting with the fire chief, you will have prevented him or her from being surprised by this information should it come to light later.

A word of caution: You must assess your organization and its leaders carefully before acting. If you are aware that your supervisor, or even the fire chief, can be retaliatory or insecure, you may decide not to bring them this information. Early discussion of the issue at hand encourages open communication; however, you must use discretion to prevent persons with a tendency to retaliate or abuse positions of authority from punishing Kevin for his personal social interactions. There are no easy answers in this scenario, and what works in one department might not work in another. This scenario and the discussion of how to handle it provide the officer with the basis to make the difficult decision about what might work best.

5. What is your last-resort action?

The last-resort action would typically be the request for a private audience with the fire chief. This step is the last resort because it escalates the matter outside the company and opens the situation up to people and decisions that are out of your control. It also puts you in the spotlight as the gatekeeper of the information and whether it is shared.

6. If you reach your last-resort action, what effect will you be dealing with two weeks after taking action?

The fact that Kevin and the city manager are becoming close socially could be considered no big deal—or it could be viewed as a subversive plot to undermine the fire chief or the fire department, even if that is not true. This is not an easy situation to handle, but the best approaches to many other issues that company officers face are neither easy nor clear-cut. You might also become the target of hostility from Kevin if he discovers that you raised the issue of his personal relationship with the fire chief. You might face sarcastic comments and a less than positive work atmosphere in your daily operations and with other members of the crew. If that negative climate develops, you may need to privately discuss with Kevin the fact that you did not create the situation; rather, he created the issue when he chose to socialize with the supervisor of the entire department.

Sensitive Policy Issues

16. Last-Minute Arrival for Shift Changes

▸ NFPA 1021: 4.2.4, 4.2.5, 4.2.6, 5.2.1

As the station lieutenant, you try to set a good example for your crew by being in the station at least 15 minutes before shift change to ensure you can be ready for work the moment the clock strikes shift change time. Ian, one of your senior fire fighters, usually does not arrive at the station until immediately before the shift change. Ian is often seen hastily slipping on and buttoning up his uniform shirt as he runs across the parking lot to enter the station just as the clock strikes the 8:00 A.M. shift change. Kirk, the off-going fire fighter in Ian's riding position, understands that shift change does not occur until 8:00 A.M., but he also knows that when the company is called for an emergency minutes before shift change, he will be responsible for responding because Ian is unlikely to be on the fire department property. Kirk, like most shift fire fighters, understands that things happen and help is needed every once in a while, whether because of a big fire response late in the shift, a fire fighter's sick child, or a flat tire on the way to work. Kirk has no issue with helping another fire fighter on occasion, but Ian's just-in-time arrival has become a habit. This pattern creates scheduling hardships for Kirk and results in overtime for the department. Kirk is discouraged by what he feels is Ian's lack of respect for him, and he looks to you to address the situation.

1. If you do nothing, what will happen?

If nothing is done, the situation will escalate because Ian's behavior will be perceived as acceptable. A bigger concern is that it may spread to others. People will typically do what is expected of them in most cases, but when the behavior does not meet the expectation and there are no consequences, a sense of inequality develops that could lead other crew members to relax their behavior as well.

2. What do you say or do?

Have a private discussion with Ian to explain the importance of being at work on time and ready to perform assigned duties the moment the shift begins.

3. With whom do you address the issue?

This issue can be handled directly with Ian. No other individuals really need to be involved at this point.

4. What are two other alternatives you could use?

The company officer has the authority to start an assignment right at reporting time. This would amplify the consequence of not being ready at the start of shift for Ian. A second alternative could be to assign the most unpleasant station chores to the last person to arrive for duty. The officer must be consistent with this assignment regardless of who is last to arrive. Do not implement such a system only on the days when Ian runs tight on time.

5. What is your last-resort action?

A documented verbal counseling session may be necessary if Ian is actually late. A word of caution: Running tight on time but reporting before the shift change time is not an offense, despite the fact that Kirk—the fire fighter waiting to be relieved and go home—might not like remaining on duty until the last minute of his assigned shift. It is not unreasonable to expect Ian to be ready for work, in uniform, with personal protective equipment placed on the apparatus, and ready to respond to a call at shift change. If that standard is met by the prescribed time for the shift change, an infraction does not really exist.

6. If you reach your last-resort action, what effect will you be dealing with two weeks after taking action?

Ideally, Ian's behavior will have changed, and he will be on time every time. Sometimes bringing the infraction to the fire fighter's attention is all you need to do to make the change in behavior.

17. Fire Fighter Abuses Sick Time Policy

▸ NFPA 1021: 4.2.5

Jeff, a crew member on your ladder company, calls in sick for his 24-hour shift. Jeff follows department procedure and notifies you and the shift scheduler two hours before shift change. Later that day, while you and the ladder company are performing commercial fire prevention inspections, you notice Jeff's pick-up truck at the bowling alley across the street from where you are conducting inspections. You are able to identify Jeff's truck because of the fire department sticker on the window and the custom graphics he installed on the tailgate. While conducting inspections, one of your fire fighters sees and alerts you to Jeff exiting the bowling alley with some other adults and children. He is holding a soft drink can and does not appear to be sick.

1. If you do nothing, what will happen?

This is where character counts. You could avoid seeing Jeff and just not worry about his behavior. After all, someone on up the chain is probably aware of this issue, and he or she will have to deal with it. If you saw Jeff's behavior, odds are that someone else on your crew or in the department saw it, too. But you are the officer, which means you have a responsibility to keep order in your department. If you do nothing, your character can be questioned. Don't allow your integrity to be compromised as a result of someone else's actions.

2. What do you say or do?

Upon his return to duty, ask Jeff to explain the situation you observed. The facts are these: Jeff was on sick leave, and Jeff was observed in public and did not appear to be sick. Ask him to explain what happened. All too often, people, including fire fighters and fire officers, take limited facts and make connections incorrectly, only to find that the situation involves more facts or other circumstances. An explanation of those facts often suffices to completely change the perception of what was observed. Ask Jeff to meet with you privately in your office or in a conference room while on duty. Listen first before doing anything else.

3. With whom do you address the issue?

Speak directly and only with Jeff, the member who called in sick.

4. What are two other alternatives you could use?

You could request a doctor's statement if your standard operating procedures or collective bargaining agreements allow for that step. This statement may provide an explanation but will also provide a layer of protection to ensure sick time was used appropriately. The doctor's statement should validate the member's absence.

Another option might be to withhold Jeff's pay if, upon investigation, it is discovered he was not really sick. This option must be thoroughly investigated and will involve other officers beyond the company officer. To withhold pay is a severe consequence that may be challenged legally, so be thorough in your part of the investigation.

5. What is your last-resort action?

If Jeff does not offer an adequate explanation for the situation you observed, you may have to resort to disciplinary action for falsification of time records and/or abuse of sick time policy.

This issue may be associated with a standard operating procedure or collective bargaining agreement that addresses appropriate steps for the officer to take directly. Review these documents and abide by them. Fire officers need to be familiar with where to go to get that information quickly and definitively.

6. If you reach your last-resort action, what effect will you be dealing with two weeks after taking action?

If Jeff has truly abused the sick time policy, he will probably be very upset. It is not unreasonable to anticipate that Jeff will attempt to tell his story to other fire personnel to garner peer support, claiming he was falsely accused. However, you could see many others who were otherwise quiet voice support for the correction of sick time abuses. Most fire personnel demonstrate honesty and integrity and exceed expectations; when others around them do not comply with expectations, they expect an officer to step in and maintain the integrity of the department. Such sentiment is not always verbalized, but it is often appreciated when poor behavior is addressed.

18. Fire Fighter Smells of Alcohol

▸ NFPA 1021: 4.2.4, 4.2.5

Rita, a member of your station ambulance crew, has been with the department for more than five years. She has previously required performance coaching sessions from officers for missed chores, minor attendance issues, and similar infractions, but there is nothing problematic in her permanent employee file. Around the station, Rita is known as an easy spirit who enjoys socializing both with other off-duty fire/EMS personnel and nonfire/EMS friends and acquaintances. This morning when Rita entered the station kitchen, some crew members noticed the smell of alcohol. One of the crew members, Bennett, a fire fighter temporarily assigned to your fire engine for the shift, sticks his head through your office door and says, "Lieutenant, you might want to check out the kitchen. Something, or some*one*, seems to smell over there."

1. If you do nothing, what will happen?

If the odor detected is caused by an abundance of alcohol actually consumed, doing nothing could put Rita, every other member in your crew, you as the officer, and any fellow member of the department who may be responding to emergencies in serious jeopardy. Even if the odor is not due to overconsumption, if you do nothing, it will erode the confidence of your crew in you as their officer because of the serious nature of the situation. The fire service has inherent risks. To add uncertainty about a member's level of impairment detracts from the ability to accomplish the mission and is unacceptable.

2. What do you say or do?

First, you must know your policies and procedures regarding substance abuse, reasonable cause for testing, processes to be used to detect the presence or absence of alcohol, and your responsibilities to report the situation to other officers in your chain of command. If you think that sounds like a lot, it is. Don't try to think your way through a situation like this without proper guidance. As the officer, you have a responsibility to either already know or quickly and discreetly consult your reference materials to make sure every step is handled according to policy. But what if there is no policy in place at the time of the infraction?

Rely on your practices in handling similar situations: Make an informed, calculated decision and act. This is not an issue to be dismissed until a policy can be developed for the future.

Generally, policies will direct you to make a personal observation, if possible, to validate or dismiss the claim before acting any further. If the odor is detectable, you generally have reasonable cause for some further testing. The next step would be to inform your superior officer. After that communication, you should inform Rita that she has detectable alcohol odor on her breath, and she is required to get an objective result to establish or dismiss any possible impairment. The likely course of action will be some type of breath, blood, or urine test but will generally involve taking the member as discreetly as possible to a testing facility or clinic. This testing will involve the member's being escorted by you or another officer to and from the testing site.

3. With whom do you address the issue?

If you personally detect the odor, you will need to inform your immediate supervisor first. Some departments may specify notification to the fire chief or chief officer, but any such notification should occur with your immediate supervisor's knowledge of that communication. Next, you will need to communicate to Rita what the specific issue is and which actions the department is preparing to take. Be discreet, direct, and open to the possibility that what appears to be may not necessarily be the case; in other words, do not pass judgment until the objective evidence is reviewed. Inform Rita that this process is intended to protect everyone involved and should be considered an opportunity to substantiate her innocence if the report of alcohol odor was erroneous.

4. What are two other alternatives you could use?

Some officers might suggest that an alternative would be to contact a law enforcement counterpart for the jurisdiction that typically interacts with the fire department and have its personnel report to the fire station to clear this matter up quietly with no one else knowing about it. If a result indicates impairment, then you can take official action. This approach might create a dilemma for the fire officer, however: What if the result is inconclusive? As the officer, how will you be able to notify your supervisor about the situation when your member insists that the results were negative?

A second option might be to allow Rita to retreat to the bunk room to rest and hope the situation will go away relatively quickly. This is not an acceptable option; indeed, it is actually worse than doing nothing because the officer has a responsibility to act. If Rita is impaired, she is a potential safety hazard to herself and to her crew, and if you ignore the situation, you are implicitly condoning the continuation of the hazard. Although this scenario involves uncomfortable choices, you must fulfill your duty to assure safe operations for your personnel and the public.

Be aware that denial will probably be Rita's first reaction to an accusation of impairment. That response may be passive or a very emotionally charged denial. Contemporary culture often seems programmed to never admit fault, for anything, anywhere, no matter how indisputable the evidence. Excuses and explanations may blame mints, toothpaste, mouthwash, deodorant, shampoo, gum, soap, clothes, and many other sources for the suspected odor. Rarely will this scenario be handled quietly despite the effort to address it discreetly. In this sense, it is comparable to the fire scene size-up of the backdraft puffing smoke upon the company's arrival at a structure fire. An officer will need to be careful to handle this situation properly the first time, because you get only one chance to deal with it correctly. The officer's size-up may be the big difference in how the potentially volatile scenario plays out, just as it would at the volatile structure fire.

5. What is your last-resort action?

Notify your supervisor that you are taking Rita off duty because of a possible safety threat to her and other responders. If you reach this point, and the department is not transporting her for evaluation, do not let Rita leave the station alone! Keep her on off-duty status in quarters or have someone drive her home, but require her to have a driver before she leaves the station.

6. If you reach your last-resort action, what effect will you be dealing with two weeks after taking action?

Tension will be high. Some crew members who do not have all the information may choose sides based on what they have heard or believe. They may think the department is maliciously targeting Rita. Conversely, you may begin to hear other members come forward about the severity or duration of the problem before an officer was made aware of it. Some may decide that a member like Rita, who pulls the bigger organization down with her poor personal choices, may not be an asset to the team at all, and subsequently she may be ostracized.

19. Fire Fighter Under the Influence

▸ NFPA 1021: 4.2.5

José, a fire fighter from an engine company across town, has been detailed to your company for the next six weeks while one of your regularly assigned fire fighters is recovering from an injury. José has a reputation for both working hard and playing hard. He has been with the department for more than 10 years and has not had any disciplinary issues of which you are aware. José is rumored to have been out partying the night before, based on Facebook posts that other members of your crew are discussing. He reports for duty and disappears by moving to a remote area within the station. Anytime other crew members approach where José is working, he moves to another remote location in the station. He tries to escape to his bunk room as much as possible. As the officer, you notice that station duties do not include significant amounts of time in the bunk room, and your crew is starting to talk in your presence about how they have not seen much of José.

1. If you do nothing, what will happen?

If (and it is still *if* at this point) José is still under the influence from his activities of the night before, and those activities or their effects are having an impact on the daily activity in the fire station, doing nothing can have a significant negative impact at a multitude of levels—possible injuries to José, other crew members, or the public; damage to the organization's reputation and the officer's reputation; and more. The stakes are high, so as the company officer you must first identify *if* there is an issue in this scenario.

All too often, there is the pull of extremes in such cases. On the one hand, if the crew likes the member, they may have a tendency to dismiss the events as "probably nothing," and you may be tempted to stick your head in the sand and hope that the issue will not come up again. You may justify your inaction to yourself based on your perception that José is a good guy who may have just made a solitary bad decision. On the other hand, if the crew is not especially fond of José, their preference may be skewed toward action, expecting the officer to take every action available to ensure that the safety of the crew and the public is not jeopardized. Your other crew members may be feeding you information with an expectation that you will send José directly home, thereby sending the message to every other member of the department that this behavior will not be tolerated. Of course, when this extreme action is taken, and especially when it is guided by bias, it can be easy to overlook the possibility that José could actually be innocent. He may have been the designated driver, sipping a cola all night to make sure his friends made it home safely, and he is now simply tired. He may have a headache from unrelated factors such as pollen, cigarette smoke, tension, or noise. As an officer, you must determine the real issue and deal with it appropriately.

2. What do you say or do?

First, consider asking the person who brought this situation to your attention to write his or her statement down and give it to you. The individual may resist creating this formal statement, but as the officer, you need to have documentation about how this inquisition started. This step also tends to weed out the malicious rumor-spreader: If the member bringing this concern to you really does not know the facts but simply wants to send you down a rabbit trail after gossip, this request for documentation will have him or her backpedaling in no time. If you observed the problem first hand, skip to the next step.

Next, make a personal observation. As the officer, you have every right to talk with José. Ask him about an assignment, make an assignment, do a surprise gear or uniform inspection—whatever it takes to engage in a face-to-face interaction with José so you can make a first-hand judgment of fact versus fiction. Look for signs and symptoms of impairment—for example, odors on his breath, intentional misdirecting of exhalation, bloodshot eyes, slurred speech, or anything else that might cause you to suspect there could be a real problem. As the officer, you have the responsibility to determine the truth by separating fact from fiction.

Should you say something to José about the allegations? The best course is to be honest. If he asks, tell him a suspicion was raised and you are taking action to see whether it is valid. If he does not ask, you do not need to alarm him unnecessarily. You may find there is absolutely no basis for the complaint. You may question José directly about the events of the preceding night and his activities' effects on him this morning. If you develop any reasonable suspicion, you owe it to the community and the members of the department to take the next step to definitively either exonerate him or provide factual evidence of impairment for protective and possibly disciplinary action. Remember—as the company officer you are on a fact-finding mission, not seeking justification for any particular course of action.

3. With whom do you address the issue?

If you observe nothing suspicious, you should consider informing the complainant that you investigated the situation and found no evidence to substantiate the claim. Be careful, because if the accuser is someone with an axe to grind against José, he or she may now want to make accusations against you for not fully investigating José in an attempt to avoid a confrontation. You should thoroughly investigate and document your findings. Not all of your findings need to be shared with the complainant. In fact, very few specifics should be shared. As the company officer, you should always be professional and have your documentation in order to recall the events should questions or accusations arise.

If you do find evidence of a possible problem, you should first review your department's policies, procedures, or regulations to make sure your actions comply with those standards. Generally, such an offense will require notification to your supervising officer(s). Call them early and let them know about the situation that is unfolding. Inform them that you have already started a preliminary fact-finding mission and believe the issue merits further investigation. Generally, the supervising officers will inform you to prepare the member for some form of analysis—breath analysis, blood test, urine test, or something else. In such a case, you will need to discreetly but directly call José to meet privately with you and inform him about the concern and your need to investigate the matter. Again, be honest. You do not need to convince José that he is guilty or innocent; rather, you are simply gathering undeniable factual information to prove him innocent if there is no problem or, in the event he is impaired in some way, to keep from overreacting or underreacting to a very serious situation.

4. What are two other alternatives you could use?

The work activity option is always a viable alternative. This would include assigning labor-intensive tasks to José to see for yourself whether he is impaired. Such an activity might involve removing the supply hose from the hose bed and cleaning and reloading it in the hose bed or conducting a surprise breathing apparatus endurance drill, a hose deployment drill up a multistory stairwell, or something similar. These are legitimate work skills that could (and should) be assessed periodically. If there is a problem, you will see it surface during this activity. A word of caution: When you assign this kind of work activity, you may also be "punishing" the rest of the crew in the name of discovery. Although a labor-intensive work-related drill for any reason is preparation, not punishment, be careful not to overwork the entire crew while trying to make an assessment of one particular member.

Another option could be to take the direct route and call José into a private meeting immediately to ask him if he has a problem that would keep him from performing his duties safely and effectively. This confrontation may or may not involve reporting the rumors that are flying about. This may very well be the best option depending on the histories of all parties involved and the guidance available from departmental policies, procedures, or guidelines. Just be aware that if your direct questioning makes José feel wrongfully accused, you may have to deal with a hostile work environment until things settle down. Even if that is the outcome, you as the officer have an obligation to help work through that situation. You may need to explain that, even if the rumors are unfounded, it is important to get the factual information to clear the fire fighter's name and reputation. You can anticipate that this kind of investigation will probably not make José feel any better, but as an officer you have a duty to investigate, and you can explain to José that you will perform your duty as expected.

When considering various alternatives, fire officers should bear in mind that some people may recognize and respect that they are doing their job, but many fire fighters and others will not understand. People are not always aware of the information you initially received, the responsibility you have as the company officer, or the results of your fact-finding assessments, yet they will claim to know what you should have done. You can anticipate that José will probably become very irate, pushing you to tell him who complained, and trying to divert the process by bringing up every other infraction ever committed by any fellow crew members, by other shifts, or even the chief. José, and possibly his supporters, will be angry! He may threaten to quit on the spot; to dig in and "earn" his time off, especially if he really knows there is merit to the accusation; or drag you or other officers down with him. Be prepared for the possibility that every action you have ever taken—or more importantly, *not* taken—could be called into question. This is not a desirable place to be, but bear in mind that you did not put yourself in this position. However, you did accept the responsibility to be an officer of your department, to carry out your duties as they relate to the crew members, the community, and the department. This is where you are needed.

5. What is your last-resort action?

If you have reasonable suspicion, and you are unable to reach supervising officers or initiate definitive testing in accordance with your departmental policies, procedures, or guidelines, you may have to reassign José to a nonpublic, nonhazardous, station duty until the other conditions have been met. By taking this last-resort action, you protect the public and the members of the crew/department, and you protect José by not overreacting and causing undue loss of pay or humiliation.

6. If you reach your last-resort action, what effect will you be dealing with two weeks after taking action?

Animosity and a hunt for a rat! If you cannot definitively confirm or disprove the alleged activity, there will be a whirlwind of unknowns and speculation occurring on both sides. This outcome is not beneficial for anyone, especially the department. You must work to bring about a definitive resolution one way or the other.

20. Pornographic Material (Print) in the Station

▸ NFPA 1021: 4.2.5

Danny, one of the department's longtime volunteer fire fighters, has been bringing pornographic magazines into the fire station. He does not want to get caught with them at home, so he brings them to the station to read. As a volunteer, Danny drops by the station to ride with different crews at various times and days, so his station visits are rarely predictable. Something else that is unpredictable is where the magazines will turn up. The magazines are sometimes left out in view even after Danny has left the firehouse. This afternoon, just after the engine company has been dispatched to an emergency call on which Danny responded, you stop in the kitchen for cup of coffee and see a pornographic magazine lying open on one of the chairs.

1. If you do nothing, what will happen?

Doing nothing is a ticking time bomb! When any product has legal restrictions, such as a minimum age to purchase, and is present in a public facility, it is a problem. These materials might also trigger a member to raise an allegation of sexual harassment or a hostile work environment because the material may be offensive to another member of the company or department. The presence of such materials certainly does not contribute to any productive business of the department. Furthermore, based on the mere presence or continued presence of such materials in the fire station, it could be argued that such materials are pervasive enough to establish that the working environment normalizes the presence of the magazines. The presence of the magazines makes a statement, real or perceived, that its content is acceptable. Such claims of harassment also can carry heavy penalties to anyone associated with contributing to the environment.

Often fire fighters will refer to the fire station as a "home away from home." In reality, any fire station is supported by public taxpayer funding, so it is a public facility and not a private home. It is only a matter of time until someone from the public, a spouse, or even a child of a fire fighter wanders through the living quarters and notices the material.

2. What do you say or do?

Since Danny is the member bringing the magazine in the station, you know who it belongs to or at least who is responsible for its presence in the station. Pull Danny aside privately and quickly, and directly inform him that he is prohibited from bringing such magazines into the station. Reinforce with Danny that the fire station is where all members present need to focus on their duties for response, response readiness, and public service. Your message is not a statement of judgment, as you may personally agree or disagree with the use of the material. The important aspect to target with your message as the company officer is that the magazines are not job-related materials, and they have the potential to offend or even place the department in a perilous legal position. Consequently, they have no place in the station.

3. With whom do you address the issue?

Deal directly with Danny. Talk to him privately, and explain he is responsible for removing the materials from the facilities immediately.

4. What are two other alternatives you could use?

Some might argue that the company officer should just pick the magazine up and throw it away. Another alternative might be to hide the material in a drawer or locker. The "out of sight, out of mind" approach might be the easier route to take, but it does not really resolve the issue. You are the company officer and you are aware of the situation—and you have a responsibility to deal with it quickly and effectively.

The company officer must know that pornographic material is present, because it is always possible for the material to be left in a public place unintentionally. For example, suppose a fire fighter was on the way to another part of the building with the magazine in hand when a call comes out. Where is the magazine going? The answer is obvious: on the closest flat surface the fire fighter can find on the way to the apparatus. Will the fire fighter remember that the material is in public view when the apparatus returns? If the call was for an accidental fire alarm activation and the company is back in the fire station within 15 minutes, maybe he or she will. But can the same be assumed for a call that takes two or three hours before returning? It is doubtful that the magazine will be on anyone's mind in such a case. There is just no reason to have such materials in any fire station for any reason.

You should be prepared for a wide variety of responses. Danny may simply take the material out of the station and not bring it in again—problem solved. But it is also possible that Danny will want to keep the material and push back a bit on the officer about his freedoms in the United States, such as the freedom to bring his personal reading material to work if desired and the freedom to read without censorship or having supervisory moral approval. The key here is not to focus on your approval or disapproval of the material but rather to emphasize the distraction from the mission that the fire fighter is there to fulfill. If Danny wants to have reading material, suggest that he read a trade magazine or training text. No one can argue that such work-related materials are inappropriate.

5. What is your last-resort action?

The last-resort action in this scenario might be to inform Danny of your expectation that he will remove the materials and not bring them to the station. Explain to Danny that should anyone in the station—be it a regular co-worker, a temporarily assigned co-worker, or even a guest in the station—see the material, that person could have grounds to make a complaint of a hostile work environment or, more specifically, a sexually harassing environment. What Danny finds acceptable is not the standard for measurement but rather what anyone who may be assigned or present as a guest of the station finds acceptable or offensive that will be the basis of judgment. A fire station is not a private dwelling nor is it just another place where an individual can express personal freedoms of choice. A fire station is open to the public and must be open and respectful of other employees', volunteers', and guests' beliefs and values. If Danny fails to comply with your directive, the issue now involves insubordination, and you may have to respond to the insubordination as well as confiscate and dispose of the material directly.

6. If you reach your last-resort action, what effect will you be dealing with two weeks after taking action?

Some fire fighters might protest your position based on what they perceive to be issues of morality. Misinformation may also run through the station like wildfire, because it is not likely Danny will share the information that he had the opportunity to peacefully remove the materials without fanfare. Rather, he is more likely to report to his peers that the officer did not approve of the material and "disciplined" him because of a personal disagreement with its content. Be prepared for this erroneous perception. Hold firm to the facts and fight the good fight for the image and integrity of the department as a whole.

You are also fighting for the image and integrity of Danny, even if he does not realize it. Your actions may prevent him from experiencing the career-impacting consequences of a sexual harassment claim or an allegation of creating a hostile work environment. It is also noteworthy in this scenario that Danny is a volunteer. Whether you are a volunteer officer or a career officer in a combination department, you should recognize that volunteers can suffer consequences that affect their careers, too. A claim that Danny has created an environment of sexual harassment toward other department members could end the volunteer career for Danny, but its consequences could also carry over into Danny's full-time workplace should such an allegation be upheld at the fire department. Although what any employer might decide is actionable is impossible to predict, many employers take a stance of zero tolerance for creating such an offensive environment. A sustained charge of a hostile work environment or sexual harassment in a volunteer fire fighter capacity could result in significant negative impacts to Danny's full-time employment.

In some cases, your credibility as an officer may actually improve with your handling of the issue. In such a trying time, remember that as an officer you agreed to take on the responsibility of leadership; when you confront this type of issue, you are clearly putting the interests of the entire department ahead of the interests of the few.

How might the interests of the department be placed first? As the fire officer who is ultimately responsible for your station while you are present, take a moment to think about these possible scenarios: Danny's wife stops in to visit and sees the magazine. A local news reporter stops by the station to gather some details about the fire a few days ago and sees the magazine. The mayor or city manager stops in to discuss facility repairs that the fire chief has been advocating and sees the magazine. Will any of these people have a more esteemed view of the fire department because of what they have seen? This magazine (or any other type of such materials) can only pull the department down, not lift it up. Although some of these people may not care about the presence of these materials, their presence will certainly not garner any additional respect for the department, the firefighting profession, or the company officer who supervises the crew of that station.

21. Pornographic Material (Digital) in the Station

▸ NFPA 1021: 4.2.2, 4.2.5, 5.2.1

Leo, a three-year fire fighter and new career member, has been downloading pornographic material on his tablet computer and has on occasion rented such media to watch on the department's DVD/Blu-ray player, which is located in the living quarters of the fire station. This behavior is not a one-time occurrence. It is not uncommon for crew members' spouses and children to stop by the station to visit their loved ones while they are on duty. You readily admit that your department needs to update its rules, regulations, and policies, but at present there is no mention of Internet or Wi-Fi restrictions, or limits on audio-visual equipment usage beyond the requirement to take care of the department equipment and use good judgment.

1. If you do nothing, what will happen?

As company officer, you cannot really claim that you did not know these videos were in the station. Thus, if you do nothing, you have, in essence, condoned the situation. Doing nothing becomes a stamp of approval to some degree. This perception is important to you as the officer because if such activities are questioned or challenged, you will be considered just as responsible as Leo, the member who brought the digital media in.

2. What do you say or do?

Once you become aware of the presence of such material, it is essential that you pull Leo into a private conversation to discuss the matter. If his behavior is a direct violation of your department's policies or regulations, you must address this breach. If policy or regulation does not cover this issue, you should talk to Leo discreetly in a professional manner, explaining the importance of keeping public service workplaces open and appropriate for all segments of the public to visit at any time, including children. This discussion would apply regardless of pay status; volunteer fire fighters, just like career fire fighters, work at the fire station. Therefore, regardless of what Leo feels is acceptable to watch or participate in on his personal time, it is not appropriate for such videos to be present in the station or viewed in the workplace.

This discussion needs to avoid a judgment about right and wrong and leave your personal beliefs out of the equation. Even if you consider watching such videos acceptable, as an officer you should not allow this behavior to occur in a public building, by public servants, even in areas marked "fire department personnel only."

Some might argue that Leo's behavior is not a big deal if everyone in the station agrees to it; however, as the officer and a supervisor, you cannot allow a workplace environment to develop where people participate in controversial and potentially offensive behavior based on who is willing to speak up against it. Participatory decision making is healthy if you are dealing with any issue that would not adversely affect the public, regardless of which final decision is made. Deciding between chicken or hamburger for dinner has no adverse consequences; in contrast, viewing pornographic videos can easily create an atmosphere of sexual harassment or inappropriate workplace behavior that can have adverse consequences for everyone involved. Remember—all members of the department are representatives of the organization whether in the station or on the street, and you are the responsible officer.

3. With whom do you address the issue?

Speak directly and privately with Leo.

4. What are two other alternatives you could use?

Approach the issue from the perspective that the fire station is a public building and that children are a segment of the community who have access to fire department facilities. Risk managers might insist, "Children should not be in fire department facilities unaccompanied by a responsible adult." This is true—but how would you explain to a parent, "Please wait here while we remove our adult-only entertainment." When you think through the issue carefully, it is truly absurd to have such materials in the fire station.

Another alternative might be to take advantage of the material as an opportunity for implementing additional training. If Leo is watching adult-only materials just to fill time, replace the material with fire or emergency medical training materials that provide work-related viewing opportunities. If you choose this alternative, be sure to clarify that the pornographic material is not welcome in the fire station in the future; otherwise, you just postpone the issue to another day.

A private discussion with Leo will probably suffice to resolve this problem. If you have any rapport with Leo, he will likely respond with respect, as long as you do not try to make his choice of videos a judgment issue of right and wrong.

5. What is your last-resort action?

The progressive discipline approach is always an option. Unless your department has a specific policy regarding pornography, the progressive discipline in this case is not so much about the material itself as it is about Leo's failure to obey a lawful order given by a company officer. When you make a reasonable request to Leo as a member of your crew, it is reasonable to expect him to follow that directive. If he fails to do so, he is guilty of insubordination. As the company officer, your crew will not likely regard you as less effective because you tried to maintain a workplace atmosphere that is appropriate for all segments of the community at all times. At the same time, you should not position your action as a value-based decision that sides for or against the "right or wrong" of pornography. Stay the course by maintaining a standard of openness and response readiness rather than one based on personal preferences.

6. If you reach your last-resort action, what effect will you be dealing with two weeks after taking action?

You may face ridicule for a period of time, but it will likely be voiced behind your back, not in front of you. Some individuals may accuse you of "judging," and you may even see an influx of inappropriate videos, magazines, or cartoons into the station. Don't let this behavior sway you. All too often in the fire service culture, people remain silent about issues to avoid the ribbing, jokes, and sarcasm that are likely to ensue if they speak up. Sadly, we are shortchanging ourselves, our crews, and our future officers when we take that tack. Where do we draw the line? Help establish a culture in your organization that promotes workplace activities that are appropriate for any segment of the community to witness. Maintaining this disciplined stance is not always easy. A good way to approach such problems is to size up station issues just as you would size up emergency scene issues. Although Leo's rented pornographic material issue is not an easy one to address, you could compare the difficulty to arriving on scene at a balloon-frame construction house on fire. What you see and think you are dealing with may become dramatically different and volatile if a breath of fresh air walks into the situation.

Just as fires are dangerous by nature yet must be handled effectively to protect the community, Leo's behavior is a sensitive issue that must be dealt with to protect others. It may have legal implications, as different localities have varying laws regarding what is permitted and what is forbidden in the workplace. Common themes in such laws include age restrictions for purchasing such materials as well as workplace protections for all employees to prevent workers from being subjected to harassment of a sexual nature or hostility in the workplace in the form of repeated exposure to offensive materials. Given that fire departments have participants of various ages and differing views about what is offensive, a department could find itself in conflict with applicable laws in its jurisdiction if this issue is not handled properly. When dealing with the topics of sexual harassment or an environment where sexual overtones create an offensive and hostile work environment, society and due process have amply demonstrated zero tolerance for such behaviors. As the fire officer, you have a responsibility to protect and uphold a work environment where such behavior is not tolerated and is swiftly and directly corrected should an infraction occur.

Consider also one other possible scenario: Suppose a crew member's grandmother, a local reporter, or a lost child wandered into the station unannounced, and the rented video is playing on the large-screen TV with all of the crew present. What happens to the department's reputation? Might the visitor make a complaint that makes its way up the chain of command? You, as the company officer, will be considered the responsible party on duty in the station at the time, and you will have to explain why this event happened on your watch. Are you ready to hang your reputation as a supervisor on the protest that "They all wanted to watch, and I didn't know Maria's grandmother (or the local reporter or the lost child) would wander into the station." That attitude could lead to an investigation regarding which other behaviors you are allowing to occur on your watch. This public relations issue says nothing about your fire-ground capabilities, your organizational skills, or your success as a motivator and developer of your firefighting crew. Instead, all of those positive aspects will be overshadowed by the discovery of the activities that you allow to take place in your station. Do you really want this to be your legacy? Keep the workplace free from such materials while on your watch.

22. Fire Fighter Takes Water Resources for Personal Use

▶ NFPA 1021: 4.2.2, 4.2.3, 4.2.6, 5.2.1

Will, a 22-year veteran of the department and a qualified apparatus operator, offers to take a new fire fighter, Carol, out for driver training with the fire engine. While away from the station, Will instructs Carol, who is driving at the time, to stop at a house up the street. Will informs Carol that the house belongs to one of Will's neighbors, who needs his new swimming pool filled with water. Will directs Carol to hook the water supply hose from the fire hydrant to the fire apparatus and then fill the pool using the hose from the fire engine. Will explains that this driving and pumping exercise is all part of becoming an apparatus operator. As Will's officer, you learn of these events after the fire engine returns, when you hear Carol commenting about the event in casual conversation with another fire fighter within earshot of your office.

1. If you do nothing, what will happen?

If no one calls the station to complain, probably nothing will happen. You may be asking yourself, "What is this point of this scenario?" When you size up a structure fire, how much attention do you give to a pile of rubble burning in the front yard after a wall collapsed? Very little; the fire is burning and it could catch something else on fire ... eventually. At the moment, though, that fire is not really an issue in need of your immediate attention. The same holds true here. The issue is probably worth mentioning, but if you do nothing, there is probably very little real hazard in terms of ongoing problems. As an officer, you must recognize when to let sleeping dogs lie.

2. What do you say or do?

It is probably a good idea to mention to Will and Carol, at some point in the future, that fire fighters have to remember we are trusted with the water resources for emergency use, not for private use for which the water distributor would normally expect payment. Some jurisdictions must report training water consumption, emergency firefighting water consumption, and other usage, so the water purveyor can account for the water resource.

3. With whom do you address the issue?

This issue can be handled privately by talking with Will and Carol, or you may choose to mention it over the kitchen table at a meal so everyone understands the potential consequences of this action. If you are not making this event a punishable or even mild offense, then it qualifies as a learning opportunity, and sharing your expectations for the future with everyone becomes a learning opportunity. If you choose the kitchen table route, you must be certain that you are not taking any disciplinary action over the incident. If you cannot rule out that possibility, do not bring up the incident with the entire crew until the discipline issue has been worked through.

4. What are two other alternatives you could use?

You can inform your crew that you expect them to discuss their plans with you before they decide to take this kind of action, so you can provide guidance. Focus on the positives: The members were performing driver training, and they were also performing pump or water supply training.

Another alternative might be to investigate with the water purveyor the availability of a water meter to be employed by the fire department so the water purveyor receives compensation for the water, and the fire department still gets the training opportunity. This option may not be completely realistic, because fire departments do not meter their fire responses, but it still allows for pump or water supply practice as well as driver training.

The second alternative, in which the fire department utilizes a water meter when filling the pool with the apparatus, provides a three-way win. The water purveyor, the fire department, and the pool owner all win, because their interests are addressed. Granted, the pool owner may want an even bigger win in which he or she does not pay for the water, but that takes away the win for the water purveyor. The pool owner still gets a rapid fill, and fire department personnel get practice stretching the hose and training on fire streams.

Additionally, Will, and possibly Carol, may be embarrassed because they were "caught." If Will did not tell you about the filling detail when the pair were heading out for driver training, you may be dealing with a more serious situation. If Will is embarrassed, you can anticipate that his response will be defensive, attempting to justify why the activity was not a big deal or even trying to draw your attention to other "minor thefts" that have occurred. Should such redirection take place, you have a bigger issue to deal with, because his response reveals that Will has at least some awareness that the action was inappropriate and could be viewed as theft. Will may think that this event should be viewed as nothing to be concerned about, but as the officer, you are responsible for any actions taken by your crew. Just as you are responsible for both members of your crew, so Will is responsible for the actions of Carol while training her. Will needs to understand that using water for any reason other than emergency responses must not result in anyone's personal gain. Although training is gained through the experience, the result of a filled swimming pool is certainly an advantage to the property owner and most assuredly not an emergency firefighting response. Will must recognize how the pool might have been filled if the fire department did not perform this service and which, if any, of those options would have required payment for the water used. Consequently, his action would be considered theft. Advise Will that if he had provided you with advance notice of such details, you might have been able to share this perspective beforehand and avoid the creation of such a precarious situation. Water purveyors are in the business of providing water for sale to the public.

Often water purveyors will allow fire departments to access the water system and use water for firefighting purposes at no charge to the fire department or the property owner experiencing the fire event. These allowances are generally recognized as a public good, with the aim of reducing a property owner's loss rather than providing that individual with a gain. When the fire department fills a pool as occurred in this scenario, it is not reducing a loss nor is it providing a benefit to the public at large. Since water that was otherwise billable was not billed in this training exercise, the activity would be classified as theft of a commodity.

There are opportunities to conduct necessary training, and even assist property owners with the proper procedures for filling a pool, without linking the two items together. Let Will know that you are not abandoning him over a decision that might be made differently next time, and emphasize that advance notice to the company officer of the entire training regimen could prevent this situation in the future. You might consider notifying your supervisor and the water purveyor of the events and the training nature of the activity, but be prepared to deal with a financial impact to either the fire department or the property owner if you take this approach.

5. What is your last-resort action?

You could inform your crew that in the future there will be no more pool-filling details, no matter when or where they may be or who requests the service. This kind of directive puts a stop to the potential controversy, but it can also create a firestorm of distrust and overreaction if the department has a history of performing such pool details. Fire fighters rally around fire fighters, and a change such as a decision to no longer fill pools after a past practice allowed it may potentially result in the membership feeling as if you are accusing Will of theft. This policy change will have a name and face associated with it, even if informally, and the officer may very well be viewed as not trusting the actions or judgment of Will in his training exercise with Carol. Realistically, change is rarely easy, and change to a practice that was often overlooked years ago (even if it was wrong then, too) is sometimes difficult to accept. But times change, and finances are often scrutinized more closely today. When a water purveyor is in the business of supplying water for consumption at dollars per gallon or donating it for emergencies, the water always comes at a price that is paid by someone. Getting your fire fighters to embrace this perspective may be difficult. They may see the fire officer as the reason for the change, even though your goal is to preserve the integrity of your crew, you as the fire officer, and the department.

Training opportunities come in all shapes and sizes. This event could easily be questioned by a member of the public who saw a fire apparatus filling someone's pool. In this scenario, the activity was a fill detail for Will's neighbor. This kind of personal link can create touchy problems. You and your department cannot afford to create the public perception that fire fighters perform favors for certain members of the public. Once you have talked with Will, Carol, and your entire crew about the situation, you can explain to any inquiring community member that the observed situation involved a new apparatus operator being trained on water movement operations but that changes have been made so that future training activities will be handled differently.

6. If you reach your last-resort action, what effect will you be dealing with two weeks after taking action?

Factions may form within your crew or even within your station or department, siding with you or against you. With either faction, the focus will be on you as the officer and your reaction to the event rather than the issue of pool filling. Be careful. In size-up terminology, don't use your master stream on smoke.

23. Fire Fighter Wishes to Unionize

▸ NFPA 1021: 4.2.5

You are the company officer of a combination fire department that is 90% career and 10% volunteer. The department has three stations and is growing; just eight years ago, the department was an all-volunteer force. One member of your engine company crew, Rick, is very vocal about forming a labor union local. He is constantly trying to gather support to put the unionization up to a membership vote. He is also pushing you, as the company officer, to join the cause, which includes asking for the elimination of the remaining volunteer positions and demanding better pay, benefits, managed work, and staffing needs. Rick is so focused and vocal about these issues and unionization that you are finding it difficult to direct Rick's work at the company level.

1. If you do nothing, what will happen?

The volatility will escalate. If you do nothing, the voices will get louder until you acknowledge them. As Rick's company officer, it is important to acknowledge Rick, so that he knows you hear him. Consider how you might influence the situation positively for all concerned.

2. What do you say or do?

First and foremost, recognize that you are dealing with an issue larger than yourself and your company. This fact means you must also study up on the rights to organize in your state. You have a great opportunity to lead your fire fighters by showing Rick the legal boundaries and processes. It is also important to be attuned to the climate in your locality, the department, the department leadership, and the community leadership. This kind of understanding is not attained simply to encourage Rick to be silent but rather to educate him of the risks and benefits, as well as the processes and limitations for what he is proposing. The mention of risk does not imply hostility or retaliation for Rick's desire to organize but rather refers to the concept of recognizing the positives of the current situation and the consideration that some of those positives may be at risk if officials believe they have to go on the defensive. The result may not be the outcome the fire fighters desire, but often people on both sides of a situation such as this become angry, defensive, and jaded about the process.

Another point for Rick to consider is the stances of his peers. If the majority of his peers do not want unionization, Rick must respect the majority decision. You can counsel him to not "force" the unionization issue at every opportunity if the membership has already spoken.

You can also educate Rick regarding the process for assessing his peers' interest in forming a labor union. Obviously, union officials will be ready to do this as well, but as Rick's company officer, your offer to help in this way makes him aware that you are not simply trying to hinder his efforts. You can control the rhetoric somewhat by offering objective assistance. This approach also sets the stage for keeping emotions in check, because if Rick fails to follow the constructive path you present to him, he is disobeying your orders, and you have a course of action to enforce an objective, constructive process. There is no need to let the pitch become fevered and volatile if there is a more constructive alternative to action. Your company officer leadership can keep the tone objective and constructive for everyone in the department.

3. With whom do you address the issue?

First, talk with Rick and explain to him that you hear him. Discuss the next steps he can take under the current laws and regulations in your jurisdiction. Second, communicate up the chain of command that a fire fighter in your company is trying to organize a local labor union. It is important for you to share this valuable information with your organization. Even if you would ultimately be included in the unionized group, you still have an obligation to your department as an officer to inform your supervisor of this potential influence on the organization. Be transparent regarding your communications and the steps you have taken to that point. You should voice your personal belief about the union, for or against, only when it is time to vote or if someone specifically asks you for your personal opinion. You are an officer and must maintain objectivity and impartiality regarding department issues.

4. What are two other alternatives you could use?

You could encourage Rick to draft a list of the pros and cons of forming a labor union local. This step seems very elementary, but as part of the assignment, instruct him to ensure he lists at least three pros and three cons. The cons he identifies will force him to think about the real opposition to his proposal.

A second alternative might be to have Rick describe his perfect scenario in writing. Ask him, "If everything went according to the perfect plan you envision, what would the labor union look like? Which issues would you address? Which structure would you have in place? Which objectives would the union adopt?" Writing the responses to these questions does two things: It forces Rick to think about the assignment more deeply than just rattling off an uninformed and superficial opinion, and it gives you an opportunity to review his responses and assess how realistic, or altruistic, his visions are. For example, Rick might envision the labor union as a vehicle to improve pay for fire fighters by 30 or 40 percent. After you have reviewed his responses, you may be able to work through unrealistic expectations to at least ground his aspirations in reality. Although increased wages may be achievable, it is not realistic to think that any municipality will devote all of its revenues to any particular department; there may be some fluctuations, but it is highly unlikely that the city's garbage collection operation will be terminated so the city can improve fire fighters' salaries, no matter how underpaid the latter may believe they are. Your follow-up discussions with Rick should not be intended to berate or discourage him; rather, your goal should be to help him identify blind spots and take precautions so he does not innocently walk into a "backdraft."

5. What is your last-resort action?

You might consider looking for a similarly sized department that has been through the unionization issue and seek their advice. This is a last-resort action, because it publicizes that unionization discussions are occurring in your department. Moreover, although size or composition of fire departments may be similar, their attitudes, personalities, and community feelings toward unions may be very different. As a result, such a consultation may give you a false sense about what to expect or how to address the issue. If you keep those variables in mind, however, you may gather some very useful information about what works and what doesn't when helping your fire fighter understand timing of unionization drives, open communications, constructive coalitions, and so on. Also be aware that you may be implicated as part of "the movement" if you seek outside information. Officers in your chain of command will certainly be interested to discover how actively you are participating with the process. Assess your personal situation and the climate of your organization to see if you need to inform your supervisor(s) of what you are doing if you reach this last-resort action, because this action involves a more active role than the other alternatives discussed up to this point.

6. If you reach your last-resort action, what effect will you be dealing with two weeks after taking action?

Lots of research! Good information does not come quickly. You may be dealing with an impatient fire fighter in your crew, a worried supervising officer or fire chief, and a potential firestorm of fire fighters for or against the issue looking to use your research as an indicator of target placement. Remember, your words carry the weight and authority of the rank you hold. Even though this is not a rank issue, you cannot separate your position from it. Be prepared to take some heat, no matter where you stand on unionization. When you size up a fire that has a heavy volume of fire showing, no matter which information you perceive and act upon, there will be others who would choose a different course of action. On significant fire events, there are many purchase points that are equally acceptable; you will pick yours and go to work. Likewise, there are many perspectives to choose from regarding how you can handle the unionization issue; pick one and charge!

The National Labor Relations Act and individual states' right to work laws address many aspects of organizing labor unions. Such laws must be respected by employers. As a fire officer, you are an agent of the employer, so you are advised to become at least somewhat familiar with such laws and restrictions. Providing Rick with some alternative perspectives as well as some thought about the potential consequences of his ideas, both good and bad, does not constitute interference, but as a fire officer you should recognize where those general boundaries have been established by law.

24. Fire Fighter Self-Discloses an Arrest

▸ NFPA 1021: 4.2.4, 4.2.5

It is shift change on a Saturday morning. One of your crew members, Brian, a fire fighter and move-up apparatus operator, approaches you. Brian says that he was arrested for an alleged assault the prior Thursday. He posted bond and was released, but he will be making court appearances soon. Brian tells you that the allegations are false; he says that his girlfriend called the police and conjured up false claims about him assaulting her. Brian tells you he needs to be at work for the income and the distraction from the current events.

1. If you do nothing, what will happen?

It would be unrealistic to expect that word of Brian's arrest will not spread. If you do nothing, you leave yourself open to accusations of a cover-up. Department leadership, community leadership, and the media may all start asking many questions once they discover that a fire fighter was arrested. Your veracity may be questioned if they discover that you knew about Brian's arrest but failed to communicate this fact to the proper department personnel in the proper method and time frame. It is important to be supportive of the fire fighters whom you supervise, but as an officer of the department, you have a duty to communicate pertinent information to the organization's decision makers.

2. What do you say or do?

In a word, communicate. Discreetly and professionally ask Brian about the details and facts as they have occurred to that point. Where was he, who was involved, what happened, when did the events that led to the arrest take place, and when did the arrest actually occur? You can avoid the "why" questions at this point; if an arrest has occurred, both sides will surely debate why the situation happened. You have a duty to relay Brian's accounting up your chain of command immediately and through a face-to-face discussion with your immediate supervisor if possible. You must handle this situation with absolute discretion. You cannot afford to provide grist for the rumor mill because someone overheard something you said.

3. With whom do you address the issue?

Talk candidly with Brian. Inform him that the department may need to make some decisions about his work status or assignment pending the outcome of any legal action by the courts. Explain to Brian that any such action is not intended to pass judgment on his guilt or innocence but rather to isolate the situation from the fire department as a whole.

Next, communicate directly with your supervisor to relay all details of which you are aware at the moment. Explain to your supervisor that more information may be forthcoming as time passes, but assure him or her that you have answered all questions you can at this point. Allow your supervisor to make the decision regarding who else might need to be notified within the fire department.

4. What are two other alternatives you could use?

You could investigate the report yourself. Examination of published arrest records might indicate whether Brian's report is true or if there is more to the story than Brian has told you. If you choose to check arrest records, you may have to explain why you are checking, and this action could raise someone's attention. In other words, the person who provides you with the arrest information is now aware that a fire fighter has been arrested before your chain of command is aware of that fact. Perhaps this is a case of "Which came first, the chicken or the egg?" but proceed cautiously and know the people in your chain of command. If you are not sure whether you will be able to inform your supervisor and the department's chain of command before they find out from someone else, act to inform your chain of command, or at least your immediate supervisor, before taking any other action.

A second alternative might be to assign Brian so that he is always with other crew members until disposition of his case within the department can be decided. Do not allow a fire fighter who has been arrested for assault to be the lone caregiver in the rear of the ambulance or part of a two-person crew on another apparatus where the fire fighter might be alone with a member of the public. Explain to Brian that this assignment does not represent a judgment of guilt or innocence but rather reflects your need to perform your duty by protecting the integrity of the entire department and the community while he allows the legal system to sort through the facts of the situation. If Brian is innocent, the temporary job reassignment will make no difference; if he is guilty, the temporary assignment can allow him to continue working with a watchful eye always present to protect the public and the department.

5. What is your last-resort action?

Communicate the situation up the chain of command yourself. This issue needs to be brought to the attention of the fire chief and possibly his or her bosses. If your supervisor is unavailable for any reason, consider moving the information further up the chain of command on your own. You cannot afford to have this information "break" in the media before your officers have been informed about the situation.

6. If you reach your last-resort action, what effect will you be dealing with two weeks after taking action?

Your crew may express hurt feelings or animosity because they may not understand the gravity of the situation. If Brian's charges are dropped or dismissed, or if he is exonerated, you may find yourself the victim of ill will owing to the perception that you "overreacted." You may also sense resentment from Brian for communicating the information outside the company level. Your direct supervisor may be upset if you passed over him or her because he or she was unavailable. In any of these cases, take some comfort in knowing that you fulfilled your responsibility as a supervising officer. Doing the right thing may not always feel good or earn you a pat on the back. To use a fire-ground analogy, you may be in charge of the incident that created a parking lot, but if no one was injured in the making of that parking lot, one could argue that you did your job well.

25. Driver/Operator with Suspended Driver's License

▸ NFPA 1021: 4.2.2, 4.2.4, 4.2.5

Tim, a fire fighter in your crew, approaches you during morning chores while you are both outside behind the station. Tim tells you that he overheard Steve, the fill-in driver/operator who is driving the fire apparatus today, say he has a suspended driver's license. While he was checking his breathing apparatus at the start of the shift, Tim heard Steve laughing with a couple of fire fighters going off-shift about the fact that eventually he would have to obtain a hardship exception for his suspended driver's license. Steve also boasted that as long as he was not pulled over, he would be fine. Tim did not agree with that view, and he wanted to make sure you are aware of the situation, since Steve is detailed to your company to drive for the shift.

1. If you do nothing, what will happen?

Doing nothing would be a very serious legal misstep and potentially create a hazard to your crew's safety. If the apparatus never leaves the station, one might argue that there would be no consequence. More realistically, you as the company officer are responsible for readiness, crew safety, departmental policy, and regulation enforcement. No fire department condones driving emergency apparatus on a suspended license, so you have an obligation to address this issue. Before diving in, make sure you get the facts right. This issue could be the equivalent of a dryer vent fire being reported as a working structure fire, so be sure to size it up.

2. What do you say or do?

First, get the facts regarding how Tim "overheard" this allegation. You must make sure his report is not a ploy to get Steve in trouble without any basis or that the facts Tim reported were not third- or fourth-hand information. Even though more distance between Tim and the real source would not negate your responsibility to investigate, having some context to address the issue intelligently may be helpful. If you cannot gather all the background information, do not assume everything is fine. You must go to the real source, Steve, and ask the obvious question directly; based on his direct answer, you can then determine what to do next.

3. With whom do you address the issue?

Ask Tim for the specific time and location when he overheard this conversation and who else (fire personnel and non-fire personnel) was present when the statement was made. This gives you some leads to follow if further investigation is needed or if Steve denies making the statements.

Next, immediately ask to speak to Steve privately. Briefly but directly explain your duty to ensure the safety of the crew, enforce the department's regulations, and uphold state and national laws. Inform Steve that it has been brought to your attention that he may have a suspended driver's license. Ask him directly: "Steve, is your driver's license suspended?" Allow him to answer. If Steve tells you his license is not suspended, request proof that he has a valid driver's license in his possession, and make a copy of it if possible. Such documentation provides airtight proof that Steve has a valid license and that you investigated the allegation and obtained evidence that exonerated the driver.

If Steve confirms that his license is suspended, you must immediately reassign him from a driving position and inform your supervisors of your actions. Consult your department regulations to determine if there is an obligation to report such driving status changes, and determine whether Steve has violated that regulation. You might also consider disciplinary action against Steve for accepting the driving assignment without informing you, the assigning officer, of his driving status. Such a responsibility to report rests with the individual, because an officer has reasonable cause to believe any fire fighter with a driver's license will maintain such a qualification.

4. What are two other alternatives you could use?

If you have a good relationship with your local law enforcement agency, you might call them to check the status of Steve's license. Be honest with the law enforcement officer; advise him or her that you heard someone may have a suspended license and that you need to ensure you do not assign someone to drive the apparatus who is legally prohibited from operating a vehicle. You might consider checking the status of the driver's licenses of both Steve and the alternate who might be immediately assigned to fill his position. This proactive step avoids disclosing exactly whose status you are questioning if the law enforcement agency reports both names have valid, unrestricted driver's licenses. It also keeps you from becoming a target of animosity among your crew for fulfilling your duty.

A second, though less effective, alternative would be to simply make an assignment change immediately with your crew and remove Steve from the fill-in driver responsibilities. This approach would have the same end result in terms of your crew's safety and your immediate compliance with the law, but it leaves many questions unanswered. If questions surface later, and your superior officers discover that you made a change in assignments, albeit an appropriate one, but failed to further investigate this serious allegation, your credibility as an officer will be brought into question. As an officer, you are an agent of the department. Treat this issue as a backdraft: When you see puffing thick smoke, don't open the front door until you have taken the appropriate precautions!

5. What is your last-resort action?

If your investigation confirms that Steve's license is suspended, the last-resort action might be to relieve Steve of all duties until his driver's license is restored. If you take such an action, make sure Steve has a ride home; do not let him depart the station in his personal vehicle. Immediately inform your supervisor. Some departments may not allow company officers to take such an action, but even if reporting through the chain of command is required, a recommendation to suspend a fire fighter from duty would be a last-resort action that would be initiated from your recommendation.

6. If you reach your last-resort action, what effect will you be dealing with two weeks after taking action?

Some members of the department may believe that you have harmed Steve by meting out such a severe punishment. Many will sympathize with Steve's financial woes and perceive that you failed to support a fellow fire fighter. As an officer, you must focus on the reality that the suspension of Steve's license and his failure to report that status change were the factors that triggered the situation and that the responsibility for both rests squarely on Steve. Rarely do innocent people have their driving privileges arbitrarily suspended. You can also take advantage of this opportunity to reinforce with your crew the importance of abiding by the law and department regulations regarding reporting. Additionally, you have an opportunity to emphasize the significant responsibility of being an apparatus operator and to reiterate that the safety of the crew rests in the driver's hands any time the apparatus rolls out of the station. This is a big responsibility that should never be taken lightly.

This issue, because it involves driving laws and licensure, brings up some unique legal perspectives for consideration. If Steve drives the apparatus without a valid license, his action is an infraction of a state law. This point is not open for discussion, negotiation, or interpretation—the courts hold that responsibility. If a crew member's driver's license has been suspended, the issue is beyond any fire department decision. Legal issues may arise in regard to a local policy or ordinance, but they play a secondary role to the dictates of state and federal law. Do not take the allegations made in this scenario lightly. Inform your supervisors of your findings; such communications might include early notification of the legal representation of the department, so they can assist in determining the most effective disposition.

26. Fire Fighter Violates No-Tobacco-Use Policy

▸ NFPA 1021: 4.2.5

Your department has a strict "no tobacco use" policy. In the policy, smokeless tobacco and smoking tobacco are treated the same. One of your crew members reported that someone on your crew is sneaking chewing-type tobacco into the building and is chewing discreetly in violation of the policy. The crew member making the report wants to remain anonymous and refuses to identify the culprit.

1. If you do nothing, what will happen?

Two things will likely happen if you do nothing. First, the tobacco violation will continue. It is unlikely the culprit will suddenly have an epiphany and just stop violating the policy. Second, you will lose credibility as an officer. If you are responsible for enforcing department policy, which you are, and you do not address a violation that has been brought to your attention, the whistleblower will lose confidence in you, and that attitude can quickly spread to other fire fighters. If you selectively enforce department regulations, you communicate to your crew that some policies are acceptable to disobey, and they will likely start testing you to find out which ones you consider important. If that happens, you will have created a lot of work for yourself, and you still lost credibility. Do not let that happen.

2. What do you say or do?

Gather as much information as you can from the fire fighter who brought the issue to your attention. When you address the issue, the more specifics you have, the greater they will aid you in sorting fact from fiction as you investigate further. This step also avoids the need to go back to the reporting fire fighter for more information after you begin investigating the allegation, which will help keep the whistleblower anonymous. Maintaining his or her confidentiality is important because if crew members believe they will be put on display for reporting problems or violations, fire fighters will simply stop reporting problems to you and leave you to discover them on your own. As you may already be aware, it is amazing how "straight and narrow" fire fighters can be when an officer is present. This scenario does not involve anyone's likes or dislikes of people or behavior but rather deals with a departmental policy violation—nothing more, nothing less.

Your first step is to identify which fire fighter is using the tobacco products. Ideally, the complainant fire fighter will have provided enough information that your investigation can start very close to the fire fighter who is allegedly using the tobacco products. If not, make increased observations of your crew to determine which member may be using tobacco on duty.

Next, a discussion with the fire fighter who is using tobacco is in order. You may start with a very brief sentence or two summarizing your department's tobacco-use policy, such as where individuals can and cannot use tobacco products of any kind. Then ask the fire fighter if he or she would like to say anything. Allow the fire fighter to confess the infraction without addressing questions such as "Who told you?" or "Why am I being picked on?" Have the fire fighter verbalize his or her understanding of the department's policy on tobacco use to ensure understanding.

3. With whom do you address the issue?

Speak privately with the reporting fire fighter first, and then meet with the fire fighter who is allegedly using tobacco. Remember—these are allegations unless you observe otherwise.

4. What are two other alternatives you could use?

One alternative is to inform the fire fighter who brought the issue to your attention that you appreciate the report and will investigate the matter further. Then make a point to observe your crew to determine if the allegation is true. You may have to intentionally ask questions or interact with members to obtain accurate information. If you observe the infraction first hand, you can act as needed by correcting the situation, which includes stopping the behavior immediately and following your department's progressive discipline steps. Although some might consider tobacco use to be a trivial matter, the real issue in this scenario is enforcement of department policy. Fire fighters must understand that policies apply to everyone and will be enforced. In case of bad policy, or at least what your fire fighters perceive to be bad policy, there are ways to propose changes. Decisions to amend or delete policies are likely to be made by officers outside the company level.

A second alternative might be to hold a surprise company drill such as a simulated response, with self-contained breathing apparatus (SCBA) and on air, while remaining present the entire time. You then can observe the fire fighters. If the rule breaker spits out the tobacco, you can follow the steps identified earlier. If the fire fighter does not spit the tobacco out, he or she will likely regret this decision at some point in the exercise. The resulting discomfort, coupled with your subsequent discussion with the fire fighter, will communicate the importance of following the tobacco-use policy.

5. What is your last-resort action?

The last-resort action might include asking other crew members if they are aware of anyone using tobacco in violation of department policy. This choice is the last resort because it involves more personnel than just the reporting fire fighter and the violating fire fighter. If the allegations are true, then more than one person can probably inform you of the details. If others deny any such violation, they may start to speculate why you are asking and why you are "picking on" the crew.

A different last-resort option might be completing the disciplinary paperwork for the tobacco-use violation and bringing the suspected fire fighter in and privately counseling him or her about breaking policy. This approach is somewhat a gamble: If the fire fighter has been violating policy, you have addressed the problem directly, swiftly, and definitively. Conversely, if the whistleblower was wrong, you violated the trust of the alleged offender, and he or she now believes you will act on any information against a crew member, even if it turns out to be false.

6. If you reach your last-resort action, what effect will you be dealing with two weeks after taking action?

Your crew will be on pins and needles for a while, wondering who is watching them. The rumor mill may stir the pot of discontent, encouraging members to believe that they must "watch their backs" or that there is a "witch hunt" in progress. You will have to work hard to provide positive activities through training or constructive department activities to restore the crew's unity and keep the focus off the naysayers. This is not easy, but you can do it.

Attitude Issues

27. Fire Fighters Critical of Fire Department Decisions

▸ NFPA 1021: 4.2.2, 5.2.1

You are a member of a four-person ladder company. Roger is your senior fire fighter and Brett is your apparatus operator. Roger and Brett complain constantly about almost every decision made in the company and the department. They grumble during morning house chores, during meals, and while returning from calls or assigned work details. Sometimes they complain about company activities or priorities of work assignments. They express displeasure with department decisions, such as changes in response areas or company assignments at active fire alarm responses. Brett and Roger seem to feel that they have a better approach to every department issue and decision, and they are quick to comment when they think a decision, plan, or activity is "stupid." They tend to make their comments to each other, usually within earshot of others in the firehouse, so they cannot be corrected for direct insubordination. When any crew member offers support or positive reinforcement regarding the company or department, Brett and Roger scold that crew member for his or her failure to recognize senseless decisions. Brett and Roger appear to be entertained by their negativity and seem to enjoy expressing their dislike for the department overall, stemming from the decisions of the company, shift, and chief leadership. Their negativity overshadows any positive thoughts or deeds, and it is depressing and frustrating the other crew members.

1. If you do nothing, what will happen?

This is when you earn your rank. If you do not address this issue, a negative attitude will certainly grow that will, over time, pervade the entire crew, then the shift, then the department. If you do nothing, you allow such sentiment to take on a self-sustaining life. In fact, if you choose to do nothing, you will, by default, give your passive stamp of approval to such negativity. Don't let this happen!

2. What do you say or do?

First, set the tone for your crew and put a stop to such comments. If the comments are made in front of others in the department, speak up immediately, and remind all personnel present that no one is forcing them to remain employed in the department. If things are truly that bad, they are always free to choose to do something else. Be careful with such words, as they may be interpreted as a threat. Such a statement is not intended to force out naysayers, but it is a reminder that fire fighters, including Brett and Roger specifically, are not victims. They have choices completely within their control. When they disagree with their organization (or even with their company/crew) on an issue, they are not powerless, but they will not be permitted to drag everyone down with their tantrums. Be ready to explain that there are proper ways to voice concerns, make suggestions, and even carry out tasks that personnel may disagree with, but fire fighters should address their concerns in a constructive manner rather than seek to spread discontent. If you make these comments in front of other members, follow such public comments up with a relatively quick request to speak to the complainers privately.

3. With whom do you address the issue?

When meeting with Brett and Roger privately, be ready to ask questions and listen intently. Talk with them (individually or together) and ask them why they joined the department. Listen for positive attributes and, if possible, key in on those positive attributes that still exist. Then ask what has changed, specifically, to cause them such displeasure. Ask them if there are reasons for these changes, and what these reasons are. Force them to really think through the sources of their discontent in your presence. Don't allow quick, arbitrary answers such as "Because the chiefs want to mess with us." Ideally, this step will encourage Brett and Roger to come to an independent realization that there may be good reasons for the things they do not like. Simply telling Brett or Roger that the status quo is the way it is or referencing a standard operating procedure as the answer to justify existing practices will not convince them; instead, you have to help them to see the reasons on their terms. This is no easy task. If they complain about policies that address safety concerns, such as a ban against riding the tailboard as in "the good old days," carefully talk them through the choices they have available, including retirement, resignation, possibly a leave of absence, or any number of other options. Reinforce the sentiment that you, the department, and the community do not want to lose them as fire fighters, but they have to assess the reality of the organizational landscape and decide whether they really want to remain a part of it. That choice rests with them.

Brett and Roger may have some valid reasons for their unhappiness, and the department may very well need to make changes in leadership, adjust policies, or fix organizational issues. If this is the case, proceed to question number 4.

4. What are two other alternatives you could use?

Meet with Brett and Roger individually to hear them out in a private setting. The key is to conduct a receptive meeting where you listen intently. Listen for underlying issues, for triggers, and for specifics that may help identify the root causes of their discontent.

A second alternative would be to have Brett and Roger write down their complaints and provide their perspective on how to best address these issues. Get their ideas regarding "If I were chief for a day," and carefully review and assess their answers. This may be the first step to identifying needed changes. It also may require Brett or Roger to put real effort into describing their feelings and the reasons for those feelings. You may be able to plant the seeds of a more positive attitude by requiring them to end their written comments with a list of everything that they find positive and enjoyable about being part of the department. Don't let them slide by providing just one or two items; make them really face the fact that there are good things about where they are.

Encourage a one-on-one dialogue with Brett and Roger. This option reduces the chance of grandstanding and provides the best opportunity for you to see what is really going on with Brett and Roger individually and together. Experience teaches us that a good person-to-person sit-down discussion about issues is very helpful for all parties involved.

5. What is your last-resort action?

Realistically, you may be thinking you should just help Brett and/or Roger get a transfer to another station, another shift, or another department. This desire is not unreasonable, but that outcome will not address the problem; instead, it will simply extend the agony and force a peer officer to struggle through the issue. Fix the problem while it is in front of you. If necessary, consider drawing up a template letter of resignation; if Brett or Roger is still disgruntled at the follow-up meeting, offer the letter for him to review, sign, and submit to the commanding officer. Again, this action must be done carefully, because it should not be your intent to just get rid of someone; rather, your intent should be to provide real, viable choices to the member for his future.

6. If you reach your last-resort action, what effect will you be dealing with two weeks after taking action?

Be ready for rumors that you tried to get rid of Brett and Roger. Others both inside and outside the department will scrutinize you and your actions. You should not let this skepticism deter you from your mission of holding Brett and Roger accountable. On a positive note, you may find the meeting and the resultant actions were refreshing to Brett, Roger, or both, and the negativity may decrease or disappear completely. Of course, some people will react by continuing to be negative and complain. Even so, you can rest assured that you have taken actions that loudly proclaim that complaints, gripes, and dislikes will be heard but not welcomed or harbored in your crew.

28. Volunteer Fire Fighters Compare Incentive Stipend Checks

▸ NFPA 1021: 4.2.4, 4.2.5

Volunteer members of your department receive stipend checks as incentives to participate in firefighting activities. Check amounts are based on levels of activity and training over a period of time, so members receive different amounts depending on their individual participation. Emma has been a volunteer fire fighter with your department for two years and recently received a volunteer incentive stipend check. While talking with Seth, a volunteer fire fighter who has been with the department for seven years, after a training one evening, Emma learned that Seth received a larger incentive check. Emma states in the presence of you and other crew members that she feels cheated and unappreciated. Although Seth is an involved member, Emma believes she is more active than he is, attending training at the fire station almost every week and going on many of the department's response calls. Seth is satisfied with his current check amount and does not want anything to change. As far as you are aware, no other volunteers are comparing incentive stipend checks.

1. If you do nothing, what will happen?

The issue grows! If you do nothing when you know someone feels cheated about something, the atmosphere at the department will certainly become a powder keg. If you are aware of the problem, get out in front of it if possible. Doing nothing here is very dangerous: If negative feelings are left unaddressed, they will breed misinformation, misunderstanding, and discontent, which can be a powerful force to have to overcome after it takes root. All too often, comparisons are directed toward other people who are in what is perceived to be a better situation without considering other people who are actually in less fortunate conditions, which would counter the claim of unfairness. In this situation, doing nothing would be similar to a fire fighter detecting a small fire and ignoring it, only to be surprised that it grows exponentially into a raging inferno requiring a significant firefight.

2. What do you say or do?

Ask to speak with Emma privately. Ask her to explain specifically what she is upset about and which facts she has to offer to assess the situation. Be careful not to defend or attempt to correct any misstatements she may make. If she is comparing her incentive/stipend with Seth's remuneration, ask how she became aware of the amount of Seth's check. Did someone share documentation, did she hear talk, or did she hear second- or third-hand rumors regarding what others were paid? Be sure to fully explore Emma's perspective on the situation. This sets the stage for further investigation if it is warranted. It is possible that Emma has been misinformed and she has taken this error as fact. This may come to light with or without your having to investigate further.

3. With whom do you address the issue?

The initial discussion needs to be between you and Emma. Don't overreact and make a mountain out of a molehill, but be ready to dig deeper if a true inequality has occurred, either intentionally or unintentionally. This may require further individual conversations with those with whom Emma is comparing her stipend/incentive payment. Any of these conversations would be like the initial discussion—a fact-finding mission to determine whether or not an error has occurred.

4. What are two other alternatives you could use?

One alternative that you may be tempted to employ is to refocus Emma on the fact that the incentive/stipend is a "thank you" for service to the department and community and should not be a source of comparison regarding who gets how much. If you are tempted to remind Emma of this fact, please reconsider. It is most likely that the real issue is the inequality concern, not the amount of money received. If you try to focus on anything other than the inequality issue, you will merely reinforce Emma's anger.

You could ask Emma to share any documentation she has that might help you recalculate the amount of her stipend/incentive check to determine whether a calculation error was made on her check before comparing her check to anyone else's. This helps set the stage for truly investigating her concerns and shows good-faith effort that you will not just dismiss her concerns.

The second alternative may be to inform Emma that you will take her concerns further up the chain of command, since this involves the department's finances. Just as a fire in a confined space or chemical facility requires specialized teams and equipment for a successful response, some issues discovered at the company level may require some initial response action that is sufficient to control the situation until specialists can be dispatched for assistance. Size up your station situations carefully because you, as the company officer, may not be able to handle every scenario on your own, even if you are willing to do so.

5. What is your last-resort action?

You may have to issue a stop-loss order for Emma and those under your command. In other words, if you cannot remedy the situation quickly and within the initial group, it is possible that Emma will continue to gripe about the situation while it is being investigated, which will make things worse, fast! In fact, any significant delays may exacerbate Emma's attitude and you, as the officer, may have to decree that the topic is off limits for discussion in and around the firehouse. If you reach this point, you should explain that the directive is intended to keep everyone focused on emergency services tasks. Reassure Emma and the rest of the crew that the issue of inequality is being addressed, but until the answers are communicated to the appropriate persons at the appropriate time, this topic is closed.

6. If you reach your last-resort action, what effect will you be dealing with two weeks after taking action?

If you reach your last-resort action, Emma and others on your crew may become angry, and they may display potential hostility toward you for attempting to limit individual expression. In such a case, it might be advantageous to assign other tasks such as training, drills, apparatus spring cleaning, and similar tasks that will help force the crew members to refocus, giving them less time to sit by idly and think about the situation. There are no guarantees that this approach will work as intended, but it is a noble effort given what you can control as the company officer.

29. Fire Fighter Critical of Pay Grade System

▸ NFPA 1021: 4.2.4, 4.2.5, 5.2.3

Juan, a crew member in your water tender company, is considering a promotion opportunity. Based on the published job announcement and the current department policy related to pay rates, he compares the pay rate for the next rank up, sergeant, against his current pay rate as a water tender apparatus operator. With his time in the department, Juan's seniority pay would not carry over to his new position if he were promoted to company officer. In fact, his additional pay for all the company officer responsibilities would be about $500per year, with only slight increases over time. Also, his opportunity for overtime would decrease, resulting in a larger financial hit in take-home pay. Juan wants to be promoted, but he is upset by the design of the department's pay grade system, and he feels as if he cannot pursue the promotion if he wants to maintain his current standard of living. Frustrated by these findings, Juan has been talking to the other crew members throughout the department about what he perceives as the department's unfair pay system for promotions.

1. If you do nothing, what will happen?

Although some fire fighters may dismiss the complaint as someone having a bad day or consider Juan to be a chronic complainer, this issue has the potential to lead to a negative culture if you do not address it.

2. What do you say or do?

You might hold a private conversation with Juan, but you also could consider an inclusive company-level conversation with part or all of your crew. In this instance, Juan is not upset about a disciplinary issue or a conflict with another individual; rather, his complaint deals with the larger system. Even if a specific person is named in the comparison, keep the conversation open and general. You should be quick to point out that no individual really has the ability to define his or her pay rate, as it is determined by the system. Hear Juan out and push him to give specific details about what he sees as unfair about the system.

3. With whom do you address the issue?

You could engage in a casual discussion outside the office in a less formal setting, with one or many crew members present. It would probably be a good idea to limit the discussion to the company at this point. If a battalion chief or other station officers or members are present, it may be a good idea to delay the conversation until just the local unit is present. This restriction is appropriate because when you deal with perceptions of unfair treatment, you are rarely able to adequately address the issue in a brief 10-minute conversation. Depending on the circumstances, a traveling fire fighter or a visiting officer might not be around to hear the follow-up discussions held hours or even days later. It is not a good idea to have anyone be privy to only partial discussions, because such incomplete communication opens the door to speculation from those parties with knowledge of only part of the story. Also, the higher up the chain of command you go with these initial discussions, the more resistance or defensiveness you are likely to encounter for even asking about the policy. Higher ranking officers may perceive that they are being accused of receiving more pay unfairly. As humans, we tend to defend ourselves when we believe we are under attack. Do not contribute to setting the stage for volatility.

4. What are two other alternatives you could use?

One alternative is to offer Juan the opportunity to submit his ideas for redesigning the system. Let him detail, on paper, what the entire compensation package should look like in his opinion, including the job responsibilities for each position. After all, compensation is bestowed for job performance, so make sure his perception of what is fair is well defined. Juan may offer vague disclaimers, such as "I don't know all that stuff; I just think Officer X should be paid less" or "I don't care about that stuff; I just think I should get more for what I do." Don't let Juan off the hook; make him think about the situation—the whole situation. The whole situation includes the requirement that the total compensation under the redesigned plan must be equal to or less than the total money currently budgeted for payroll.

As another alternative, you might ask Juan to conduct a survey to assess what other departments pay their varying ranks. Make sure to include comparable demographic samples and similar regions. A small rural combination department will not have a pay structure similar to that found in a large city such as Miami, New York City, or Chicago. You might even ask Juan to find at least three departments that provide lower pay rates at all ranks as well as at least three that have higher pay rates. Otherwise, Juan, like most people, will tend to look only at those fire fighters who are better off than he is without acknowledging the fire fighters in comparable circumstances and departments who are paid less, have older equipment, and have more difficult working conditions than he does. Those people would love to have what Juan has. Yes, your organization is the "well-off" department to someone out there. Help Juan recognize this fact.

5. What is your last-resort action?

If the alternatives given earlier do not seem to quell the perceptions of inequality, and Juan remains vocal about the unfair pay system, you may have to consider a private conversation with him about whether he wants to continue as a member of the organization. If the pay structure of your department really is not in line with those of similar companies, this topic is definitely worthy of discussion. In contrast, if the system is deemed to be fair in light of factual comparison data, then Juan needs to consider the desirability of his ongoing employment in the organization. You should not issue a threat on this score, nor should your conversation veer toward retaliation for any reason. If the system has been evaluated and will not change, Juan's continued complaints will just make the culture of the organization more negative. You cannot afford to allow this deterioration to happen. As the officer, you must encourage Juan to do some soul searching and either (1) get on board with the current system and stop complaining or (2) consider leaving for greener pastures, meaning a department whose pay system is more in line with his expectations.

6. If you reach your last-resort action, what effect will you be dealing with two weeks after taking action?

Ideally, Juan will decide to accept the current system. If not, you must take stronger control of the situation by directing Juan to stop any workplace discussion of this topic. If you reach this point, be prepared to react should Juan fail to follow your orders. Juan would then be guilty of insubordination or failure to follow orders, and this issue no longer relates to perceptions of the pay scale.

30. Fire Fighter Critical of Fire Department Leadership

▸ NFPA 1021: 4.2.4, 4.2.5, 4.4.4, 5.2.1

Michael has been a fire fighter/EMT in your department for more than six years. Prior to joining your department, he volunteered at other fire departments, both in the region and in other states. Michael is very critical of the department's administration and voices his criticism to members in your crew, members of the shifts before and after yours, and even members of other departments, if given the opportunity. He believes that the administration's priorities are misguided and that it tries to get along with other city departments when it should instead fight for the fire department regardless of the opinions of those in city hall. Michael believes that every leader at headquarters is out of touch with the fire fighters on the line who actually provide the service and that every decision the current administration makes and every policy it drafts reveal the fire administrators do not know what they are doing. He claims that they fail to select the right company leadership, purchase the right apparatus, and provide the right equipment and tools, and he insists that their training plans are outdated and "playing it too safe." Michael is quick to back up his claims with stories of other departments he has worked in, in which the administration "got it." "This place,"

Michael maintains, "doesn't have a clue." You have overheard partial conversations between Michael and other fire fighters on your crew when Michael was expressing his criticism of and disrespect for the department's administration, and some of your fire fighters have repeated some of Michael's statements directly to you.

1. If you do nothing, what will happen?

If you do nothing despite your awareness of this issue, you are agreeing with Michael by default. Your failure to say or do anything communicates such tacit agreement. Some might argue that just because an officer doesn't say anything, it doesn't mean the officer agrees with Michael. As the officer, you are aware that Michael—who is your charge, your subordinate—is upset with the actions of the department's administration. The key points to consider here are that your organization, represented by the administration (your bosses), is being attacked—maybe rightfully, maybe wrongfully—but as an officer, you cannot allow your subordinate to constantly criticize the organization. There are many ways to objectively debate the merits of actions, decisions, or issues, but a member's expressions of being upset about something in many cases may be the testing ground for finding those who agree with the member's opinion. All it takes is one person to take sides and verbalize that agreement: "You're right, that *is* upsetting; they *did* make a bad call on that one." Such agreement signals that the seeds of negativity are planted and their effects are spreading. You have a duty to dig in and explore this issue further. Should you decide to do nothing, you allow the fires of discontent to grow and send the message to Michael and the others that you agree with them.

2. What do you say or do?

Ask Michael to talk with you privately about the specifics regarding who, what, when, and why he is upset. Listen. Do not agree, disagree, or justify—just listen. When you think Michael is finished, probe for more by asking if there is anything else that is bothering him. You may find it best to then close that conversation and let Michael know you need some time to think about what he had to say. Ask Michael to give you a few days to figure out how you might help the situation. This achieves two objectives: First, it provides a cooling-down period for Michael. Second, you have effectively stopped the spread of the negativity. You may ultimately determine that Michael has a good reason to be upset, but you cannot reinforce his emotion right now.

3. With whom do you address the issue?

Talk directly with Michael. If more than one crew member is upset, talk with the members individually. Just as group brainstorming works to create energy in a problem-solving setting, so the synergies created in a group complaining of feeling wronged can become an increasing challenge that you may not have the authority to work through completely on your own. Any statement you make to one member may easily be misinterpreted by another, and any attempt to clarify or explain yourself may look like back-pedaling to some, especially if some members already distrust the department officers.

4. What are two other alternatives you could use?

One alternative is to request that Michael write up his complaint and submit it to you, his officer. Asking someone to write the details down often clarifies "gripes" and distinguishes them from real issues. This exercise will also force Michael to think about his thoughts and feelings as he writes them down. This assignment may make the issue over which Michael was originally upset not seem as bad as it once appeared to him. As the officer, you should not allow Michael to simply internalize his complaints and pretend externally as if nothing is wrong in an effort to avoid the writing assignment. Don't be fooled by an apparent change of heart, because Michael may actually be fuming on the inside, possibly even more upset than before.

A second alternative might be to have Michael engage in an exercise in which he gives the administration the benefit of the doubt. Ask Michael to humor you for a moment and think of scenarios where what appears to be so upsetting to him might not be the reality at all. For example, if Michael is upset about a new policy, ask him to consider the possibility that the policy change is based on a new labor law of which he is not aware. What if the fire department administration does not really want to issue this policy but is getting pressure from city hall, or a citizen complaint, or numerous citizen complaints—would this factor change his perceptions? Perhaps what he observes is not the whole story, and the whole story is not really open for disclosure or discussion. Then ask Michael, "What would you do if you were in command and the issues about which you are upset were really out of your control?" Don't let Michael respond with vague or overly simple answers: "If we don't have enough money, then we should get more money." That approach will not work at a fire scene, either; just because the fire department shows up, the officer or the fire fighter cannot just order the fire to go out and assume that it will happen.

5. What is your last-resort action?

As a last resort, you can remind Michael that an organization is only as effective as its team. If Michael is so upset with the administration that he cannot work through the issue, he must realize and accept that it is not up to subordinate personnel to question superior officers or to make decisions that are outside the scope of their positional responsibilities. As Americans, we have the freedom of choice: We can choose to stay with an organization, or we can choose to leave it for another organization or a different vocation altogether. What we cannot choose to do as individuals is to single-handedly change the mission, the direction, or the decisions of the greater organization. The fire fighter always has the freedom of choice. If he is so terribly upset about the administration or the department, Michael's choice is to either remain with organization and support its administration, policies, and practices or to leave the organization.

6. If you reach your last-resort action, what effect will you be dealing with two weeks after taking action?

Michael may think you are trying to push him out of the department. Remind him that he is not a victim and that he has a choice regarding his continued involvement in the department. Ideally, this reality check will refocus Michael's energies and allow him to see that he has productive options to stay and contribute; in contrast, to stay and gripe about everything not to his liking is not an option available to him.

It is important to note how powerful this issue is, especially for the company officer. The company officer has the ability to probe and recommend action to Michael without posing a threat. For example, the last-resort action sounds a bit like "love it or leave it," but the company officer in most organizations does not possess the power or responsibility to fire someone or to make determinations about membership or employment status. If you choose to elevate the issue and bring a higher ranking officer into the situation, that dynamic changes a little. If you have investigated this issue with a chief officer present, the "love it or leave it" options really could have some "leave it" authority, which creates a threatening environment rather than a coaching environment. The battalion chief or chief may be a great person, truly interested in solving the root issue, but his or her mere involvement escalates the stakes, and could be perceived by Michael as a threat rather than an opportunity to work through what is upsetting him. This is where company officers' actions and conversations can truly make or break a department!

31. Fire Fighter Critical of City Administration

▸ NFPA 1021: 4.2.4, 4.2.5, 4.4.4, 5.2.1

Sam, a 14-year veteran of the department, is very vocal to the rest of his crew and fire fighters on other shifts about his dislike of the mayor and city council. He believes that these officials do not value the fire department, citing decisions the city has made against pay raises and additional staffing and its failure to support the replacement of needed fire equipment and apparatus as examples of their disregard. When other fire fighters point to positive actions within the fire department, Sam gives credit to the fire chief and fire administration but continues to criticize city officials for their failure to care about or provide for services to the citizens. Sam is so distressed by his perception of the mayor and city council that it is the only topic he wants to discuss. You are Sam's company officer and hear his criticisms more and more often throughout the shift, over multiple shifts, and with increasing intensity. You dismissed Sam's early commentary as healthy venting of frustration, believing it would release the tension and Sam would let the issue go; now, however, you observe that Sam's rants continue and appear to be an increasing obsession.

1. If you do nothing, what will happen?

If you do nothing, you risk having Sam spread a culture of negativity in the crew, then in the station, and ultimately throughout the department. In this scenario, the member is upset with a person or group that is not his direct employer. Every person has the right to form his or her own independent opinions about people and representation, especially elected representation. With that said, if you do nothing, especially if Sam is upset about something related to the fire department, you are taking a very dangerous gamble. Although the city officials with whom Sam is upset are not part of the direct employing organization (that is, the fire department), the mayor and city hall staff still oversee or direct the various city departments. In essence, if you allow Sam's criticism to continue, his behavior could be compared to the adage of "biting the hand that feeds you." As the officer, you should cautiously engage in the situation.

2. What do you say or do?

Speak to Sam privately, and seek to understand why he is so upset by the issue(s). Listen carefully, assessing whether there are any points you can clarify and identifying which issues you or Sam may be able to influence. For example, if Sam is upset because the mayor is spending significant amounts of time trying to bring a large company into the city, and Sam does not think that is time well spent, or if he does not like the company with which the city is engaging, these issues are really not open to influence by any fire fighter or the fire department.

In contrast, if Sam is upset about an issue such as the city manager's decision to "brown out" fire stations to save money, then that is an issue with which the fire department is involved. In such a case, the department might potentially be able to exert some legitimate influence—but even then the influence is unlikely to involve Sam as an individual.

For issues where Sam's distress is rooted in an area that he or the department has no real ability to influence, short of engaging voters in a general election, then it becomes important to direct Sam to keep his opinions about the situation to himself while at work or around the station. Inform him that this directive to keep his views of city hall to himself is not an effort to invalidate his opinions but rather reflects the reality that expressing such discontent with an official has no place in his chain of command.

3. With whom do you address the issue?

Obviously, you need to have this conversation with Sam directly. However, as the officer, you might want to seize on this opportunity to communicate the same message to the other members of your crew during a casual conversation. A general statement acknowledging the reality of differing opinions but directing your subordinates to keep their political or personal opinions to themselves while at work (career or volunteer, as both work at the fire station) would be appropriate. If fire fighters feel an overwhelming need to express such discontent, they are perfectly free to do so when on their personal time away from the fire station. At the same time, remind all of the members that as fire fighters they are always "on" to some degree. Caution members not to speak out too freely despite their strong emotions, because they cannot completely disassociate themselves from the fact that they are members of the fire department. A fire fighter's strong emotional outburst may create a backlash for the department that is unintended and unwanted.

4. What are two other alternatives you could use?

One alternative is, after listening to Sam and assessing whether the root cause of his discontent is a contemporary issue, to seek a meeting between Sam and the offending official, with the assistance of your chain of command. Seek to provide a controlled opportunity for Sam to express his opinions and empty his baggage, if that is possible. You may need to issue a disclaimer to the official that Sam's views do not necessarily match the views of the fire department, but emphasize that as an organization that supports its members as well as the city or parent organization, this meeting is important to the long-range well-being for the parties involved. This alternative must be considered with full understanding of the nature of all individuals involved. If Sam has a short temper, is a chronic complainer, or tends to focus primarily on "what is in it for me," this alternative may not be advisable. The same holds true if the official who is the object of Sam's ire has any of these personality traits. Proceed with caution. Just as an emergency medical call dispatch at a street fight calls for a heightened sense of awareness and some safety precautions, so this type of station issue and this alternative to address it call for heightened caution. Plan for contingencies in case the original plan to address Sam's negativity starts to break down.

A second alternative might be to have Sam write a letter to the person or board with which he is upset. Ensure that this communication uses a professional tone representative of the profession of firefighting. Make sure that Sam understands this letter is to be submitted to you and only you—no copies, no sharing the letter with others. You as Sam's officer should be the only recipient, but the letter should be addressed to the offending official. This alternative forces Sam to take constructive action, clarify his thoughts and put them in writing, and potentially cool down. After you review the letter Sam submitted, if you determine it merits further action, you can bring that up to your supervising officer.

5. What is your last-resort action?

This issue, because it does not involve the fire department operation directly, may call for consulting your supervising officer to discuss options. Generally, the scenarios in this text are intended to force you to think and develop your own decision-making options so as to increase your skills and abilities. Nevertheless, it is important to realize that some issues are so unusual or so loosely connected to the operations of the fire department that they merit the "two heads are better than one" approach. A second officer—a second tier of rank—might at least discuss the matter with Sam. Again, options to correct this situation may not always exist, so the method of dealing with it becomes the focus.

6. If you reach your last-resort action, what effect will you be dealing with two weeks after taking action?

Be careful that you do not send an unintended message to other fire fighters, in your crew or otherwise, that if they are unhappy and complain vociferously enough, you or other officers will automatically take up their cause. Your involvement is not any indication of agreement or disagreement with Sam's issue; rather, you are taking action to deal with a situation developing in your station, on your watch, with your personnel, that must be reasonably addressed to ensure it does not lead to bigger or more volatile subsequent issues.

32. Layoffs Affect Company Morale

▸ NFPA 1021: 4.2.3, 4.2.4, 5.2.1

City budget cuts have reduced your company staffing from four crew members to three. Over the six years they worked together, the crew of four had developed into an effective response company and became close friends. The remaining three members are upset about having lost a colleague. From their expressions, kitchen table chatter, and lack of enthusiasm for their tasks, you know they are angry, sad, and discouraged. They are also concerned that, with the reduced staffing, emergency response tasks will take longer to complete and they may even risk injury while rushing to do so much in the first few critical minutes of a response. There are times when their discouragement turns into outright anger over the circumstances that brought about the reduction in staffing.

1. If you do nothing, what will happen?

If you do nothing, the crew will remain upset. Although you may not have any direct control over staffing levels of companies or the department as a whole, you cannot allow the members of your crew to remain upset. You have options to constructively get through what has been placed before you.

2. What do you say or do?

The first step is to drain the emotion. Next, focus on the constructive, not the destructive; that is, refocus the crew on their responsibilities. Finally, keep an eye on things to make sure emotions do not start to simmer again undetected. Just as you would use a thermal imaging camera to detect hidden hotspots to complete fire extinguishment, so you are responsible for assessing and detecting the hidden hotspots with your crew and doing something about them.

3. With whom do you address the issue?

The initial discussion should involve the member or members who are vocal about their feelings in this matter. The follow-up discussion would likely include all members of your crew. It is important to recognize that, after the emotion is drained somewhat, as the company officer, your authority has only a limited reach in ensuring the crew's attitudes are properly focused. Any attempts you make to communicate your expectations for the entire department membership will be futile. Make sure your company is squared away and has balanced their release of frustration with a properly focused attitude regarding the job in front of them, and encourage them to remain upbeat as they successfully complete the tasks of that job.

It might be wise to have a separate, private conversation with your supervisor, the battalion chief, or the shift commander to at least keep them informed of the climate of your crew, the emotions and background, and the steps you have taken to keep things under control and focused on positive outcomes for the department's service delivery. This communication should not be an effort to "snitch" or to request your supervisor's help but rather the provision of a situation report on an issue that could become volatile quickly for the shift or the department.

4. What are two other alternatives you could use?

Ask your crew to write a report summarizing their concerns. In creating this report, they should research data to support their concerns (if such data are available). Instruct them to also find data that would support the decision to reduce staffing from four to three members for a company. This research will require the members to look at the current situation from another perspective. They may disagree with that position, and they may present data that support their position rather than the staffing cuts, but make sure the crew do their homework. Give them a deadline for completing the report, and direct them to submit it to you alone; they should not publish it or give it to the media or other members of the department. Their final report should be in writing, rather than in the form of a verbal report. This assignment is important for many reasons. First, they cannot ignore your request and just let it fade into the background. Second, the assignment helps them develop skills for future officer positions within the department—writing skills, critical thinking skills, and even the skill of learning to implement policies that they may not favor personally. Third, the crew must focus on something constructive, rather than just continuing to complain and negatively affecting the fire department's culture.

A second alternative might include completing some constructive homework. Ask your three-person company to record realistic data for the amount of time necessary to perform basic rudimentary functions within your company—for example, access the hydrant, set up the pump, and stretch the initial attack line. Compare these data for your three-person company with the data for a four-person company. Do the same for other rudimentary functions: emergency medical assistance for a critical patient, a fire alarm investigation in a high rise, a vehicle extrication. Although other response companies will eventually join your company on the emergency scene, show what your company can do (and cannot realistically do) so you have real, local data that can be used to discuss the impacts to your community. If the data reveal the company's ability to perform company tasks does not differ significantly with four people versus three, be honest about that finding, and do not allow your crew to embellish the results in an effort to persuade others to take their view.

5. What is your last-resort action?

Communicate to your crew that if staffing levels remain at three people per company despite any results or data they are able to produce, you will define for the company what the rules of engagement will be for the reduced staffing. This is where the job gets rough on you, the company officer. You must recognize that you are not authorized to make decisions regarding the staffing level, and you cannot display any public disagreement with the organization and the decisions its administration makes, at least not with your peers or subordinates. As an officer, you must respect the chain of command and the operations of the department and carry out the tasks assigned as a supervising officer rather than assign blame to others. As an officer of the department, you must act as if you personally made this decision. If that statement causes you angst, remember that you voluntarily accepted the rank of officer. You agreed to support the continuity of the organization, for the good of the organization, when you accepted the promotion.

Be careful not to confuse support for the organization with support for any particular leader or fire department or city hall leadership or any particular decision those offices have made. Your attempt to quell the fire fighters' discontent and find the good in the situation is not a display of weakness or simply being agreeable to authority but rather part of being an effective leader in an organization. An officer must function as part of something bigger than himself or herself or the officer position itself.

Defining the rules of engagement with three-person staffing in the company will require some work on your part. You will need to enumerate the tasks that were assigned to the fourth crew member, who is now absent, and redirect those tasks for every type of call to which you respond. Perhaps two fire fighters stretched the attack hose line in the past while you completed the 360-degree walk-around of the building, but now one fire fighter starts to deploy the attack hose line and awaits your return to complete the task. For emergency medical calls where each of the four fire fighters previously took specific equipment inside, you may have to send someone back outside to the apparatus for a second trip to get all the needed equipment. These operations may take longer, but you have to recognize your unavoidable limitations. It is unrealistic to simply instruct one fire fighter to carry all the equipment that was previously carried by two people. Keep your expectations grounded in practicality. Discuss these changes with your crew, explaining that you do not expect them to simply do more with less and achieve the same outcomes within the same time frames as they did previously with four people in the company.

6. If you reach your last-resort action, what effect will you be dealing with two weeks after taking action?

In two weeks, you will likely continue to experience delays in service to the customer. These delays are not your fault, and they are not the fault of your crew, so it is important to keep a watchful eye to ensure your crew does not start looking for someone on whom to place blame. You must cope with the new reality, and you must overcome your limitations safely to the best of the abilities of your crew.

If you are in a department with a low volume of calls for service, it is possible that you may not observe any noticeable change after implementing the last-resort change. Some changes—even ones that seem like terrible ideas at the outset—may surprise you by not being nearly as bad as you anticipated. Encourage your fire fighters to look for pleasant surprises everywhere. Not every call will result in a poor outcome because of the reduced staffing levels. Additional staffing is always helpful and appreciated, but that does not mean reduced staffing will automatically be catastrophic.

33. Company Dissatisfied with Design of New Apparatus

▸ NFPA 1021: 5.2.1, 5.4.6

You are the company officer for Engine Company 14. A new piece of fire apparatus is assigned to your station. Your crew members do not like the compartment design of the newly received fire apparatus. They think the storage compartment located inside the cab is too big, reducing the room available for the crew, and is poorly designed, because the equipment intended to go inside that compartment could be stored more conveniently in other places. The engine is the first brand-new apparatus the station has received in years, and you believe that your crew is unappreciative. They keep finding what they perceive to be problems, such as equipment shelf locations in the apparatus body compartments, the location of switch panels for the apparatus operator, and the rear-facing seating design. They blame the seat design on the large storage compartment in the cab, which they believe is unnecessary.

1. If you do nothing, what will happen?

Possibly nothing will happen if you do not act. You and your crew will probably have to listen to the gripes and complaints for a while, but then things will eventually settle down and the issue will just fade away. The question is, do you want to rely on hoping the issue will just go away? There is another possible outcome of doing nothing: The rumblings of discontent may grow louder and stronger, and your crew members may begin to pick apart other things—maybe even you! If you do nothing, a message may also be transmitted through the grapevine within your organization that your crew is not appreciative of the new apparatus assigned to your station. New fire apparatus is typically the safest, most reliable equipment and is purchased at a great expense. You do not want your crew to be perceived as ungrateful and complaining, as this label could ultimately be attached to you, your crew, and your department.

2. What do you say or do?

Pull your crew together and remind everyone of some basic facts. First, apparatus design and procurement are not typically a crew responsibility. Although the department may (or may not) form a committee asking for input on apparatus purchases, it does not change the fact that the apparatus design and construction are not the crew's *responsibility*. Second, the apparatus is new, and there are many positive aspects of having a brand-new apparatus with which to operate. Focus on the positive features of the unit—and there will be positives—and do not let the negativity dominate the discussions.

3. With whom do you address the issue?

Initiate the discussion with the crew members who are upset. If all but one or two are complaining, it may be worth having a crew meeting to brief everyone on these points, even the crew members who have not criticized the apparatus. This kind of leadership also provides some extra benefits: It enables you to communicate to the members who are pleased with the apparatus that you will not ignore problematic situations in hopes they will fade away, and it allows you to show leadership by outlining specific expectations for what merits complaint and what does not. Generally, fire fighters appreciate a leader who takes steps to keep the complainers in line.

4. What are two other alternatives you could use?

One alternative is to instruct the fire fighters vocalizing complaints about the apparatus to put their pencil where their mouth is. Ask each fire fighter to write out his or her issues in a memo to you, outlining the specific details he or she sees as problems and suggesting at least one or two alternatives to improve on each of these issues for future purchases. This assignment should include researching possible cost implications, feature modifications, and trade-offs. For example, to create more compartment space, the apparatus may need to have a smaller water tank capacity or carry less supply or attack hose in the hose storage bed.

A second alternative, albeit one that will draw more attention to your crew's discontent, could include requesting to have the department's reserve unit sent to your station for a while. The crew could then run that unit as the primary response apparatus while the new, "poorly configured" apparatus sits in the station unused but in plain sight. This option may require some coordination because such a swap would not necessarily be implemented across all shifts—just yours. After some time passes, the benefits of the new apparatus will probably start becoming more evident to your crew members.

5. What is your last-resort action?

You can direct the complaining members to simply cease discussing the apparatus issue. If you implement this last-resort action, you must be clear about your general desire to receive input from your crew regarding issues over which you have control. On this particular issue, however, neither you nor the crew has the power to change the apparatus compartment design. Therefore, there will be no further discussion on that subject.

Some might suggest the fire fighters who are complaining about the apparatus should be detailed to another station that has older apparatus, but this is not a good idea for a few reasons. First, transferring a problem rather than dealing with it does not display effective leadership. Deal with the issue where it is: If you simply move the complaining fire fighter and his or her new company receives a new apparatus in a few years, the person will probably react in the same manner, because he or she was not taught how to constructively deal with disagreement. Second, you do not want to break up an otherwise well-functioning crew over an apparatus issue or because you as the leader are trying to make a point. Assuming they are cohesive and effective, don't let an object lesson in constructive disagreement break up an effective street response capability. Of course, if your crew was not a cohesive unit before the new apparatus arrived, you have other issues to deal with (keep reading and practicing).

6. If you reach your last-resort action, what effect will you be dealing with two weeks after taking action?

Some members of the crew may resent you for quieting their voices, even if their voices were not productive. It is also very possible that the issue may be resolved so that your company moves on productively. If your company does resolve the issue in some fashion and moves forward, you will have played an extremely important part in that outcome because you acted to create the outcome rather than because chance favored you. As an officer, you may have the opportunity to inject some positive comments in this scenario by drawing attention to other positive actions, emergency responses, or constructive feedback received from the crew members regarding other issues.

When sizing up a structure fire, you do not size up fire suppression exclusively; rather, you must consider search and rescue components, weather, and resource availability, among other factors. Don't let the apparatus issue become the sole consideration for your crew over the week or two following your directive.

34. Excessive Equipment Location Changes Request

▸ NFPA 1021: 4.2.6, 5.2.1, 5.4.6

Another station in your department has received a brand-new quint fire apparatus. Due to your station's relatively slower activity level, the quint fire apparatus that the other station had been using is reassigned to your station to finish out its life cycle within the department.

The quint is still in fair condition, but once it arrives at your station, your crew begins looking for ways to change equipment to make it more user friendly. Tara, a two-year fire fighter, comes to you almost every shift with a new idea for where to move various hand tools and how to rearrange the fans and saws. At first, you approved a few of the proposed moves, but Tara keeps coming to you about once a week with a new idea. First, she wanted to move the extinguishers out of the compartment and mount them on the side running boards. Next, she suggested moving the apparatus operator's turnout equipment from the first compartment to the rear compartment. Then she advocated adding some standpipe hose packs to the officer's side compartment, followed three weeks later by a request to move the standpipe hose packs to the driver's-side compartment because it would work better. You have heard from some of your fire fighters that your colleagues on the shift before and the shift after yours are complaining to their officers about the frequent equipment relocations. The other shifts are becoming frustrated because those personnel never seem to know where the equipment is located when they need it.

1. If you do nothing, what will happen?

Placing equipment in a location of personal preference is not necessarily linked to the availability for use or functional effectiveness of such equipment. Expression of dislike, if momentary, can be acceptable to allow crew members such as Tara to give their opinions or even vent their frustration, and you may choose to leave the issue alone. If, however, Tara's expressions are dramatic, overly angry, or directed at particular decision makers or other crews, then you need act to settle a potential people problem before it gets started.

2. What do you say or do?

Although doing nothing is a reasonable option, you must recognize your responsibility as the officer to engage in active monitoring of the situation. Do not assume comments or complaints are a case of mere venting and ignore the issue. You will have to check in often to see if the sentiments are a recurring topic of conversation for Tara or other fire fighters. This kind of monitoring can be as simple as paying attention to the topic of conversation when you walk across the apparatus floor or onto the back porch of the station or even as you are backing the apparatus into the firehouse. If the complaints fly every time your crew uses the apparatus, then it is time to say something to your crew.

What do you say? Probe those members who are unhappy—in this case, Tara. Ask her specifically what she feels is the problem. Do not ask in a confrontational style or tone but with an attitude of genuine interest. You may or may not agree with Tara's sentiments, but in either case you should attempt to gain understanding from her perspective. When she seems to be finished, probe a bit more with questions such as "Do you have any other concerns?" or "Is there anything else specifically you can think of that makes you feel this way?" Close the conversation with a statement such as "Let me think about the issues you have raised." Do not agree or disagree with Tara at this point, although it may be tempting. Let time work in your favor. While you, Tara's officer, are considering things, it is harder for her to continue complaining because you have listened to her input. Tara may be waiting for a response, but you have a better understanding of her thoughts, and ideally the fire fighters complaining will acknowledge that you care enough to ask.

3. With whom do you address the issue?

If you engage in the discussion previously outlined, do so individually with Tara or individually with the members who have expressed discontent. As an officer, you will always face the challenge of having your private conversations discussed and possibly even misinterpreted or misquoted. This hazard is magnified when such conversations are held with multiple individuals. However, when you address a group of 2, 3, 5, or 20 people, all or some of whom have known concerns, you run the risk of triggering a pack mentality. The complaint is typically vocalized by the one member or a few members who feel the strongest about the issue. It can influence or even feed the flames of discontent when you try to single-handedly gain control over the group's emotions. Rarely will you have a mixed group where discontent is strongly expressed and another member in disagreement with the vocal fire fighter speaks up just as strongly to counter the complaint. Often, the content members will remain silent in the interest of peacekeeping or maintaining the status quo. For these reasons, you should discuss the issue with the complainants individually.

4. What are two other alternatives you could use?

One alternative is to have Tara design an alternative equipment placement plan, where all equipment currently in place remains on the apparatus but is simply moved to the location she believes is more appropriate. Assign Tara to run an experiment by physically moving the equipment around to ensure it fits, is accessible, and does not require more investment, such as fabrication of additional shelves or relocation of existing electronics, to make it work. Make sure she understands that this is a finite experiment and that once equipment is relocated and assessed, and perhaps even photographed, it will be returned to the original location with the task completed before the end of the current shift. It is amazing how many times a perceived "solution" will work only if the department purchases a particular additional item or invests some additional money to. Nevertheless, the experiment may actually reveal a better way—and there is nothing wrong with discovering a better way of doing things. If Tara discovers a more efficient equipment placement plan, you should be prepared to take her findings to the appropriate individuals within the organization's chain of command to investigate the possibility of making the experimental change more permanent. Be patient. You may encounter hurdles or delayed decisions beyond the company level. Remind Tara and your crew that a delay might not necessarily indicate "No" but rather "Not right now."

The second alternative would be to put Tara's thoughts to a peer test. Assign Tara the task of surveying the other members who work on the apparatus to determine whether they feel the current equipment placement is acceptable. Tara may discover that the majority of the other fire fighters working on the apparatus are satisfied with the current setup. A little peer influence may tone her rhetoric down a bit. Conversely, if her survey reveals that other crew members and other crews are unhappy with the current equipment placement, ask Tara not just to gather the survey results but also to compile her peers' input regarding alternative setups. Again, task her with researching costs, effects, and collateral considerations and have her report her findings in a memo to you for review. If Tara's report is lacking detailed information, refer it back to her with any unanswered questions. These sorts of dialogue and documentation are good practices for your crew in looking beyond the initial emotions and teach them that as their officer, you are not the dumping ground for all of their concerns, with the expectation that you will solve every problem. It empowers your crew members to be part of the solution. The fire service is a team endeavor, and this alternative offers an opportunity to reinforce that.

5. What is your last-resort action?

After listening to Tara's input, reviewing research results provided (if research assignments were made), and assessing any directives from your supervisors that must be considered, the last-resort action would be to make and enforce a directive to Tara that this particular issue is no longer a topic open for discussion. Your decision may be that there will be no changes; alternatively, you may have forwarded the suggested changes up the chain of command for consideration. Reiterate that in the meantime and regardless of outcome, the topic is no longer up for discussion.

6. If you reach your last-resort action, what effect will you be dealing with two weeks after taking action?

If you decided no changes were needed, you will likely hear rumblings about your dictatorial leadership, see signs of discontent, and potentially face some animosity directed toward you by Tara and other fire fighters who agreed with the changes she proposed. If you decided to make recommendations for equipment placement changes to your supervisors, it is likely Tara and her peers will be impatient to receive a response or angry if the response is negative. If denial from above is the ultimate outcome, as the officer and the agent of the organization at the company level, you must find constructive points on which to focus Tara and your crew—some enjoyable training evolutions, response successes, organizational successes, or something else constructive to keep Tara and the crew positive about the organization, even when a particular decision did not turn out as hoped. Maintaining a positive attitude toward the organization is even more important now because Tara knows you supported the changes. Do not allow blame to fall anywhere within the organization. Blame is a negative cancer that eats at the very lifeblood of the greater organization. As an old saying goes, "He who throws dirt, loses ground."

SECTION 2

Battalion Chief/Shift Commander

Introduction

Up to this point, the opportunities to size up various situations have all been from the perspective of the company officer. The company officer has a tough job of keeping a company of people working together as a team. Each team member brings background, perspective, and experience that influence how the company performs and how the company officer must lead. Similarly, when you climb the organizational chain of command, you find the same issues take on different perspectives when considering the impacts to an entire shift of personnel. The battalion officer is typically charged with leading multiple company officers, just like the company officer is responsible for leading multiple individuals in the fire station. The battalion chief is less concerned with the internal details of individual fire stations or what happens on the shifts preceding or following his or her assigned shift rotation. Rather, the battalion officer is very concerned about operational readiness, a smooth and effective response, and a cohesive team with clear direction for all the fire fighters and fire officers assigned to his or her shift. Decisions made at the battalion level typically have a filter of consideration about how those decisions will affect all personnel in all stations on that shift.

Just as failing to walk 360 degrees around a fire building during size-up to consider all factors while establishing an incident action plan for a fire emergency response can have devastating consequences, so too can decisions made by the battalion chief without consideration for shift-wide implications have similarly devastating consequences for consistency, integrity, and effectiveness.

The battalion officer can have a different title based upon the size of the organization and, at times, regional influences. The title battalion chief is common, but in some departments the title might be shift commander, district chief, captain, or assistant chief. When this section makes reference to battalion chief, it is referring to any officer who typically is responsible for overseeing and leading a shift—a particular day in a schedule rotation. The battalion chief usually is directly responsible for the company-level officers assigned to him or her who might be stationed at multiple locations throughout the jurisdiction. The battalion chief is unique in the fact that this officer, despite being responsible for an entire fire department operation for a given period of time, often still has strong ties to the shift personnel due to all the nuances of working for 24 hours together. The battalion officer, like the company officer, responds with the fire companies at all times of the day and night, works holidays together with the station personnel, and can sometimes be considered by fire fighters and company officers on that shift as their advocate to fire department administration and other shifts.

Consider the fact that the very same issues discussed in the previous section regarding the company officer have the potential to become issues with which the battalion chief may need to become involved. Such involvement may be initiated because of ineffective results from a company officer's problem solving or because two or more company officers have conflicting viewpoints of how best to address an issue. The battalion officer may also become involved should peer-level battalion officers report that the company officers on different shifts have disagreements they cannot resolve.

It was mentioned in the Section 1 opener that at times officers must make decisions that do not have win-win outcomes and a decision often requires determining which parties must change their behavior. When those decisions are made, the individuals involved may wish to challenge that decision, and the battalion chief is the next level of authority in the chain of command to review such decisions. Likewise, the battalion chief must consider implications to the entire shift and even how decisions might impact the adjacent shifts when solving problems. With this in mind, some of the scenarios presented in this section will be the same scenarios presented in Section 1. The significant difference is the viewpoint. The stakeholders have changed and are typically a broad audience, and the consequences may have much wider impacts.

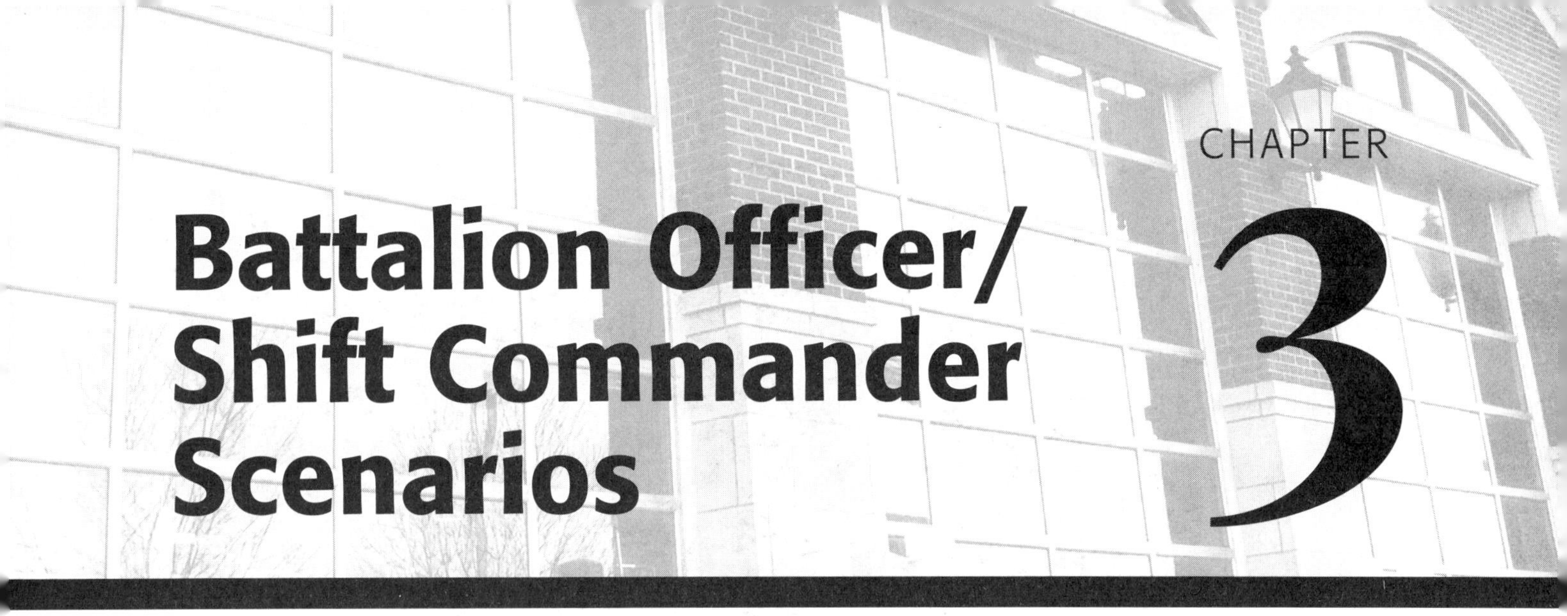

CHAPTER 3

Battalion Officer/ Shift Commander Scenarios

Work Product Issues

1. Dirty Dishes

▸ NFPA 1021: 4.2.2, 4.2.4, 4.2.6, 5.2.1

Your department expects that every individual will clean up after himself or herself. Chore lists, supervision, and history all reinforce the assumption that a fire station's kitchen, along with the rest of the station, is to be kept clean and orderly. You receive a complaint from a company officer, Lieutenant Malone, that the off-going shift routinely leaves a mess in the station kitchen. The oncoming crew arrives to find the floors unswept, the table dirty, and dirty dishes left in the sink. A discussion between your company officer, Lieutenant Malone, and the off-going company officer, Lieutenant Redding, has not resolved the issue.

1. **If you do nothing, what will happen?**
2. **What do you say or do?**
3. **With whom do you address the issue?**
4. **What are two other alternatives you could use?**
5. **What is your last-resort action?**
6. **If you reach your last-resort action, what effect will you be dealing with two weeks after taking action?**

FIGURE 3-1 It's still like this every morning! Our company officer has been told; when is something going to change?

Courtesy of Jeff Barlow.

2. Low Fuel Gauge

▸ NFPA 1021: 4.2.2, 4.2.4, 4.2.6, 5.2.1

A company officer, Captain Garcia, reports to you that the pumper is not being fueled regularly, and his shift is consistently noting a low fuel gauge during their morning apparatus checks. Captain Garcia has discussed this issue with the off-going company officer, Captain Smith, but the situation has not been corrected.

1. **If you do nothing, what will happen?**
2. **What do you say or do?**
3. **With whom do you address the issue?**
4. **What are two other alternatives you could use?**
5. **What is your last-resort action?**
6. **If you reach your last-resort action, what effect will you be dealing with two weeks after taking action?**

FIGURE 3-2 Is the fuel gauge showing ¾ full? Would you consider this gauge just above or just below the ¾ full mark?

Courtesy of Jeff Barlow.

3. Changes to Run Card Assignments

▸ NFPA 1021: 4.2.6, 5.2.1, 5.4.6, 6.2.1

One of your company officers, Lieutenant Vickers, constantly offers ideas to change the running card assignments used by the computer-aided dispatch system on the grounds of efficiency. It is not unusual to receive an occasional suggestion from a company officer for a change to a run card, but most previous suggestions have been based on new roads or road construction. These infrequent changes were relatively obvious and were consequently implemented. The changes that Lieutenant Vickers is suggesting, however, consistently create a decreased call volume for his company.

1. **If you do nothing, what will happen?**
2. **What do you say or do?**
3. **With whom do you address the issue?**
4. **What are two other alternatives you could use?**
5. **What is your last-resort action?**
6. **If you reach your last-resort action, what effect will you be dealing with two weeks after taking action?**

4. Negative Feedback from Personnel

▸ NFPA 1021: 4.2.4, 4.4.2, 5.4.5, 6.4.6

Chief Willis, the operations chief, assigns you the task of assessing morale within the department and reporting your findings to him with suggestions for improvement. As you begin to make inquiries among the companies, you sense some reluctance by the line personnel to answer your questions. You decide to supplement your interviews with an anonymous survey instrument regarding the climate of the department. You are shocked to receive an overwhelming number of responses from personnel suggesting that you are the source of the department's problems. Some believe that you are inconsistent in how you treat people and that you are selective regarding accountability. Some believe your supervision is unfair. Others report they do not trust you. The line personnel seem to have a perception that you are not proactive when it comes to department needs, such as repairs.

1. **If you do nothing, what will happen?**
2. **What do you say or do?**
3. **With whom do you address the issue?**
4. **What are two other alternatives you could use?**
5. **What is your last-resort action?**
6. **If you reach your last-resort action, what effect will you be dealing with two weeks after taking action?**

5. Rearranged Apparatus Equipment

▸ NFPA 1021: 4.2.6, 5.2.1, 5.4.6

A new ladder truck has been assigned to Station 3, and your shift's crew members at Station 3 do not like the location and arrangement of equipment within the apparatus. You were aware from making your routine station rounds that the crew of Station 3 was not in complete agreement with some of the equipment placements on the ladder truck, but this is the first time you have heard that the crew has been rearranging equipment for use during its shift. The battalion chief from the shift following yours, Battalion Chief Striker, informs you that his company officer at Station 3 has reported that his shift regularly arrives to find that the equipment on Ladder 3 has been rearranged and is not in the locations where it is supposed to be stored on the apparatus. The issue comes up daily, and the crew has strongly expressed their anger to their company officer.

1. **If you do nothing, what will happen?**
2. **What do you say or do?**
3. **With whom do you address the issue?**
4. **What are two other alternatives you could use?**
5. **What is your last-resort action?**
6. **If you reach your last-resort action, what effect will you be dealing with two weeks after taking action?**

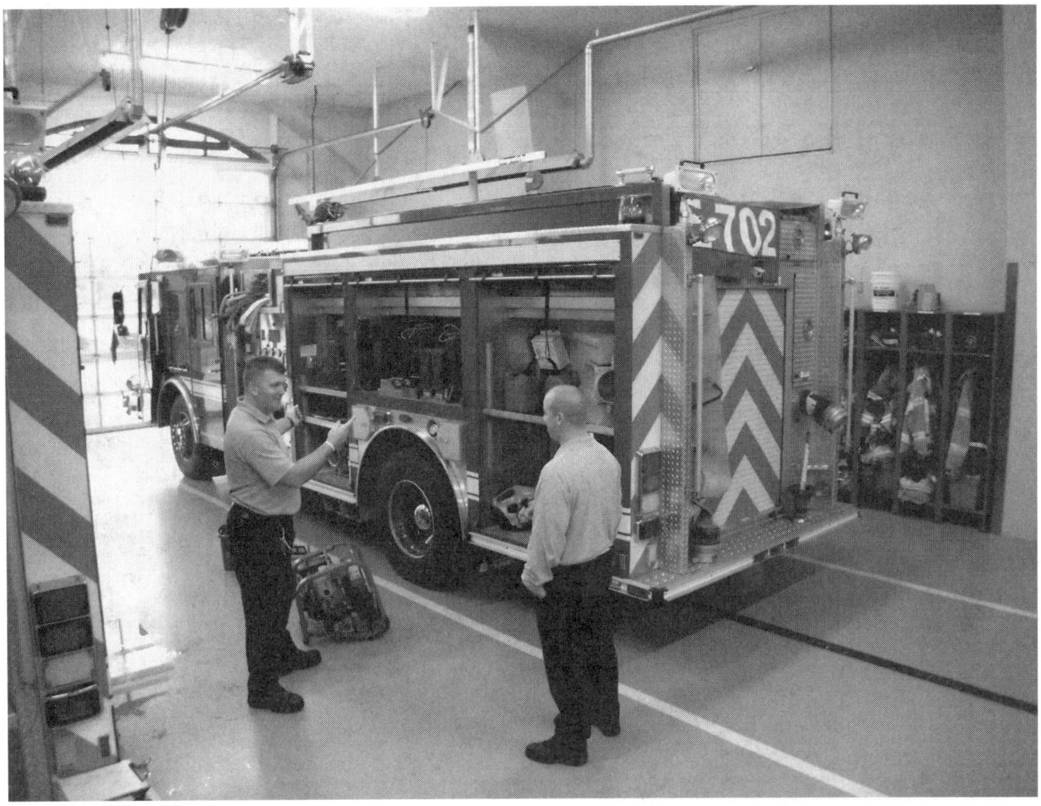

FIGURE 3-3

Courtesy of Jeff Barlow.

6. Turnout Gear Assignment

▸ NFPA 1021: 5.4.3, 5.4.5

Your deputy chief gives you the assignment to research new turnout gear options. The department historically buys this gear from what is regarded as the chief's favorite brand. After hearing the complaints of the line fire fighters following responses and training evolutions when the turnout gear is actually in use, you know that the field personnel do not really like that brand. You have to give a recommendation based on line personnel's input in one week and must determine which steps to take and how to present your research.

1. **If you do nothing, what will happen?**
2. **What do you say or do?**
3. **With whom do you address the issue?**
4. **What are two other alternatives you could use?**
5. **What is your last-resort action?**
6. **If you reach your last-resort action, what effect will you be dealing with two weeks after taking action?**

7. Citizen Complaint: Speeding Apparatus

▸ NFPA 1021: 4.3.2, 5.2.1

You receive a phone call from a member of the public. The caller reports that earlier today he saw one of your fire apparatus speeding through an intersection just as he was about to pull out onto the street. The caller says the apparatus "could have killed somebody."

FIGURE 3-4 The company officer must handle emotionally charged complaints from the public, including complaints about his or her particular crew.
Courtesy of Jeff Barlow.

1. **If you do nothing, what will happen?**
2. **What do you say or do?**
3. **With whom do you address the issue?**
4. **What are two other alternatives you could use?**
5. **What is your last-resort action?**
6. **If you reach your last-resort action, what effect will you be dealing with two weeks after taking action?**

Interpersonal Dynamics Issues

8. Inappropriate Comments to a Peer Officer

▸ NFPA 1021: 4.2.5, 5.2.1

A company officer, Lieutenant Scott, is poking fun at a fellow officer in another station, Lieutenant Stewart, regarding Lieutenant Stewart's dating practices. You have heard some of your line personnel under Lieutenant Scott's command make references to this teasing during station rounds. Some of the comments started when the two companies were together for training. During the training session, Lieutenant Scott learned through overhearing casual conversation between Lieutenant Stewart and another fire fighter that he had an upcoming date scheduled with a new acquaintance. Lieutenant Scott heard enough to begin teasing Lieutenant Stewart. Lieutenant Scott seems to find pleasure in making others squirm. He has gained a reputation of deliberately putting other crew members on the defensive over the past nine years with the department, even when he was a line fire fighter. Just three years ago, when he was still a senior fire fighter, he consistently harassed the new recruits assigned to his station. Any time the two companies are together at training or during the wrap-up of an incident response, Lieutenant Scott takes the opportunity to get a jab in at Lieutenant Stewart. You are not aware of any disciplinary actions documented in his file over the incidents, and if any of the crew members present are bold enough to say anything to Lieutenant Scott about his constant ribbing, he simply replies that he is just joking around with Lieutenant Stewart.

1. **If you do nothing, what will happen?**
2. **What do you say or do?**
3. **With whom do you address the issue?**
4. **What are two other alternatives you could use?**
5. **What is your last-resort action?**
6. **If you reach your last-resort action, what effect will you be dealing with two weeks after taking action?**

9. Inappropriate Comments to a Fire Fighter

▸ NFPA 1021: 4.2.5, 5.2.1

A company officer, Lieutenant Scott, is poking fun at Andy, a fire fighter in another station, regarding Andy's dating practices. Specifically, Lieutenant Scott comments on the women's names, hair colors, and interests and even the restaurants Andy allegedly selects for his dates. Lieutenant Scott sarcastically instructs Andy regarding how to improve on his choices in women, the venues to frequent with his dates, and ways to "land a wife." You have heard some of your line personnel under Lieutenant Scott's command refer to his remarks during station rounds. The comments started when Lieutenant Scott's company was attending training with Andy's company, and Lieutenant Scott overheard part of a casual conversation among Andy and some of Lieutenant Scott's fire fighters—enough to begin chiming in with his self-perceived expertise on dating matters for Andy.

Lieutenant Scott has a reputation for frequently finding pleasure in making other fire personnel uncomfortable. He has demonstrated this behavior throughout the past nine years with the department, even when he was a line fire fighter. Just three years ago, when he was still a senior fire fighter, he was known to constantly pick on the new recruits assigned to his station. Lieutenant Scott pokes fun when the two companies are together at training or during the wrap-up of an incident response; any time he can get a jab in at Andy, he does. If any of the crew members present are bold enough to say anything to Lieutenant Scott about his ridicule, he replies that he is just joking around with Andy.

You are the shift commander supervising Andy's lieutenant and Lieutenant Scott. You are not aware of any disciplinary action in Lieutenant Scott's file regarding the historical incidents.

FIGURE 3-5

1. **If you do nothing, what will happen?**
2. **What do you say or do?**
3. **With whom do you address the issue?**
4. **What are two other alternatives you could use?**
5. **What is your last-resort action?**
6. **If you reach your last-resort action, what effect will you be dealing with two weeks after taking action?**

10. Fire Fighter with Skill Deficiency

▸ NFPA 1021: 4.2.3, 4.2.6, 5.2.1, 5.2.2

You are a shift battalion chief in a combination department, and keeping career and volunteer members on the same page is a constant challenge. A career member, Jill, reports to you that one of the volunteer members, Dan, has not mastered basic fire fighter skills, such as breathing apparatus emergency procedures, hose stream selection and safe use, and even fire behavior recognition—all of which are necessary to perform successfully as a fire fighter. Jill informs you she does not trust Dan to work alongside her on emergency calls. She tells you that she has reported her concerns to her company officer but that nothing has been done about it, and Dan is still responding to calls.

FIGURE 3-6

1. **If you do nothing, what will happen?**
2. **What do you say or do?**
3. **With whom do you address the issue?**
4. **What are two other alternatives you could use?**
5. **What is your last-resort action?**
6. **If you reach your last-resort action, what effect will you be dealing with two weeks after taking action?**

11. Career Member Questions Volunteer Members' Skills

▸ NFPA 1021: 4.2.5, 4.2.6, 5.2.1, 5.2.2

A longtime volunteer member, Frank, files a complaint with you, the battalion chief, regarding a recently hired career member, Tony. Frank believes that Tony is arrogant and condescending, and he is concerned that Tony's attitude is negatively affecting the company. Frank's written complaint describes that Tony "struts through the bay" with the other volunteers in attendance for training and that he is often heard telling the volunteers that if they work hard, they, too, can someday be hired and "learn how to be a 'real' fire fighter—not just play fire fighter when a call comes in." Frank reports that he has discussed the issue with the station lieutenant, but nothing appears to have been done.

1. **If you do nothing, what will happen?**
2. **What do you say or do?**
3. **With whom do you address the issue?**
4. **What are two other alternatives you could use?**
5. **What is your last-resort action?**
6. **If you reach your last-resort action, what effect will you be dealing with two weeks after taking action?**

12. Negative Feedback About a Superior Officer

▸ NFPA 1021: 5.4.5, 5.4.6, 6.4.6

Chief Willis, the operations chief, assigns you the task of assessing morale within the department and reporting your findings to him, along with suggestions for improvement. He gives you no further direction, as is often the case at a battalion command level. Through your face-to-face discussions during regular station visits and brief survey of other shifts' personnel who are not readily accessible for face-to-face meetings with you, you receive overwhelming responses from personnel suggesting that Chief Willis is the source of many of the department's problems. The shift personnel perceive that Chief Willis is out of touch with the line and inconsistent in his decisions affecting the department membership. Lack of respect for and inconsistent or even unfair treatment of line personnel is becoming more commonplace. Personnel cite many specific examples to justify their feelings, so you are fairly certain the unhappiness is not just a matter of the personnel not "liking" Chief Willis. Furthermore, equipment and facilities are aging, and the line fire fighters believe that Chief Willis is not lobbying vigorously for needed improvements; they interpret this situation as Chief Willis being unappreciative of the hard work they accomplish at the street level of the department.

1. **If you do nothing, what will happen?**
2. **What do you say or do?**

3. **With whom do you address the issue?**
4. **What are two other alternatives you could use?**
5. **What is your last-resort action?**
6. **If you reach your last-resort action, what effect will you be dealing with two weeks after taking action?**

13. Company Officer Disregards Chain of Command

▸ NFPA 1021: 4.4.4, 5.2.1

Captain Biltz is an enthusiastic company officer at fire department headquarters. Whenever he has operational or administrative questions, he routinely bypasses you as the battalion chief and solicits the deputy chief officers and sometimes the fire chief for answers instead. These questions range from the rationale behind the company training rotation to which content might be covered in an upcoming promotion process so that he can be well prepared. Captain Biltz never brings these types of questions to you, his immediate superior. Sometimes Captain Biltz even reports the questions' answers back to you, as if to keep you informed. You are frustrated with his actions. Captain Biltz is a relatively new captain, holding the position for only a year now. His enthusiasm for the position and eagerness to obtain answers to his questions are how the questioning of the deputy chiefs originally started, but over the past three months, the frequency with which Captain Biltz's questions bypass you as his supervising battalion chief has increased significantly, occurring once or twice a week now.

FIGURE 3-7

1. **If you do nothing, what will happen?**
2. **What do you say or do?**
3. **With whom do you address the issue?**
4. **What are two other alternatives you could use?**
5. **What is your last-resort action?**
6. **If you reach your last-resort action, what effect will you be dealing with two weeks after taking action?**

14. Fire Fighter Fraternizes with City Manager

▸ NFPA 1021: 4.2.5, 4.4.4

Your department has a chain of command that all members understand they are expected to follow. You learn from your ladder company officer at Station 2, Lieutenant Ewing, that a member of his company, Kevin, has become increasingly friendly with the city manager. Kevin did not previously know the city manager, but since her appointment and Kevin's election as fire department labor representative, the city manager has been inviting Kevin to dinner and sporting events with her. Kevin shared this information with Lieutenant Ewing one day during casual conversation during a shift weight-lifting session. During the same conversation, Lieutenant Ewing mentioned that it was nice to have fire department personnel getting some face time with the city manager outside of city hall. Kevin then revealed that in some instances, the only city personnel in attendance at these events were the city manager and Kevin. Kevin went on to say that the city manager's family and Kevin's family had also attended multiple family picnics together. The conversation between Kevin and Lieutenant Ewing during this weight-lifting session was the first Lieutenant Ewing had heard of this interaction between Kevin and the city manager. Lieutenant Ewing recognized this relationship to be highly unusual but was not sure what to do or say about it, so he pulled you aside one day when you made station rounds to make you aware of the situation. The lieutenant further informs you that based on general comments made during the lieutenant's interaction with Kevin, he believes Kevin's off-duty conversations with the city manager sometimes involve fire department business.

1. **If you do nothing, what will happen?**
2. **What do you say or do?**
3. **With whom do you address the issue?**
4. **What are two other alternatives you could use?**
5. **What is your last-resort action?**
6. **If you reach your last-resort action, what effect will you be dealing with two weeks after taking action?**

Sensitive Policy Issues

15. Last-Minute Arrival for Shift Changes

▸ NFPA 1021: 4.2.2, 4.2.5, 4.2.6, 5.2.1, 5.2.2

As the battalion commander, you try to set a good example for your company officers by being in your work station at least 15 minutes before shift change to make sure you can be ready to go to work the moment the clock strikes shift change time. Captain Jimenez usually does not arrive at the station until immediately before the shift change. He is often seen hastily slipping on and buttoning up his uniform shirt as he walks across the parking lot to enter the station before the clock strikes the 8:00 A.M. shift change. Captain Kirk, the offgoing captain, understands that shift change does not happen until 8:00 A.M., but he also knows that when the company is called for an emergency minutes before shift change, he must respond because Captain Jimenez is unlikely to be on the fire department property at that point. Captain Kirk, like most company personnel, understands that things happen and help is needed every once in a while, whether because of a big fire response late in the shift or a flat tire on the way to work. Captain Kirk has no issue with helping another company officer on occasion, but Captain Jimenez's just-in-time arrival has become a habit. It has created scheduling hardships for Captain Kirk and resulted in overtime for the department. Furthermore, Captain Jimenez has assigned his crew to make sure everything is ready to go before he arrives at the station, including placing his personal protective equipment on the apparatus in the officer's position and having an attendance report for his company ready as he enters the door. Both Captain Kirk and many of Captain Jimenez's fire fighters are discouraged by what they feel is Captain Jimenez's lack of respect for others. One day after shift change, one of Captain Kirk's fire fighters calls you to tell you about this situation and the frequency with which it is occurring. The fire fighter informs you that Captain Kirk deserves timely relief and that his crew is concerned that although he is frustrated, he will not complain or risk creating tension with his peer. The fire fighters working for Captain Jimenez see what is happening but also are hesitant to raise the issue out of concern that it might appear insubordinate to their captain if they jump the chain of command, especially regarding an issue in which they are not directly involved. No one knows exactly what to do about the situation, so this fire fighter decided to call you to report what is occurring. They look to you as Captain Jimenez's officer to address the situation.

1. **If you do nothing, what will happen?**
2. **What do you say or do?**
3. **With whom do you address the issue?**
4. **What are two other alternatives you could use?**
5. **What is your last-resort action?**
6. **If you reach your last-resort action, what effect will you be dealing with two weeks after taking action?**

FIGURE 3-8 Being at work on time and being ready for work on time are two different things.
Courtesy of Jeff Barlow.

16. Lieutenant Under the Influence: Reported by Direct-Report Fire Fighter

▸ NFPA 1021: 4.2.4, 4.2.5, 5.2.1

Lieutenant Jones has been detailed to your shift from another shift to serve as your company officer for Engine 5 until the regularly assigned Engine 5 lieutenant recovers from an injury. Lieutenant Jones has a reputation for working hard and playing hard. She has been with the department for more than 12 years and has not had any disciplinary issues of which you are aware. Lieutenant Jones is rumored to have been out partying the night before her current shift, based on Facebook posts. Some of the fire fighters who were at the party called and posted to fellow fire fighters who are now on duty at Station 5 that Lieutenant Jones was pictured and referenced in comments that placed her at the party well into the evening. She reports for duty, assigns the daily chores, and disappears by moving to a remote area within the station. Any time other crew members approach the area where Lieutenant Jones is working, she moves to another remote location in the station—an office to catch up on paperwork or her bunk room to make her bed. A text message from an off-duty fire fighter to a fire fighter friend assigned to Engine 5 that morning mentioned the Facebook posts and observations at the party. One of the Engine 5 fire fighters decided to make a phone call to you, the battalion commander, to let you know about the Facebook posts. The fire fighter raises questions about Lieutenant Jones's pre-work activities and her continued movements away from the remainder of the Engine 5 crew personnel.

1. **If you do nothing, what will happen?**
2. **What do you say or do?**

3. **With whom do you address the issue?**
4. **What are two other alternatives you could use?**
5. **What is your last-resort action?**
6. **If you reach your last-resort action, what effect will you be dealing with two weeks after taking action?**

17. Lieutenant Under the Influence: Reported by Off-Duty Fire Fighter

▸ NFPA 1021: 4.2.4, 4.2.5, 5.2.1

You receive a phone call from Ryan, a fire fighter in the department assigned to the previous day's shift. When Ryan stopped by Station 5 earlier in the day to pick up some supplies from his locker, some crew members on duty mentioned to him that Lieutenant Jones, their officer, was rumored to have been out partying the night before reporting for duty. These rumors were based on Facebook posts and conversations with some of the fire fighters from other stations and shifts who were at the party. According to Ryan, the crew said that Lieutenant Jones reported for duty, assigned the daily chores, and disappeared to remote areas within the station. Any time other crew members approached Lieutenant Jones, she was reported to move to another remote location in the station. Ryan called you late in the afternoon because, upon further reflection throughout the day, he thought you should know about the potential issue.

1. **If you do nothing what will happen?**
2. **What do you say or do?**
3. **With whom do you address the issue?**
4. **What are two other alternatives you could use?**
5. **What is your last-resort action?**
6. **If you reach your last-resort action, what effect will you be dealing with two weeks after taking action?**

18. Pornographic Material in the Station

▸ NFPA 1021: 4.2.5, 5.2.1

You stopped by Station 11 to deliver some requested supplies from headquarters, only to find that no one else is in the station. The company had been dispatched to a fire in a neighboring district, but it was reassigned while en route to stand by in the neighboring district's fire station until some of the initial response units were released from the fire scene. While walking through the station to drop off the supplies in the kitchen, you look across the room and notice an adult pornographic magazine lying on the recliner in the TV room.

1. **If you do nothing, what will happen?**
2. **What do you say or do?**
3. **With whom do you address the issue?**

4. **What are two other alternatives you could use?**
5. **What is your last-resort action?**
6. **If you reach your last-resort action, what effect will you be dealing with two weeks after taking action?**

19. Citizen Complaint: Transport Delay

▸ NFPA 1021: 4.3.2, 5.2.1, 5.4.5

You receive an e-mail from the operations chief concerning a citizen complaint that alleges one of your ambulances, Ambulance 7, experienced navigation issues last week en route to the hospital, which resulted in delayed patient treatment. According to operations chief, the patient's family has complained that the ambulance driver did not know where he was going and that as a result of the long travel time getting the patient to the hospital, the patient suffered adverse effects, including delayed surgical intervention and spread of infection. You have asked for a report from the dispatch center regarding Ambulance 7's incident times for that particular call. A quick review of the computer-aided dispatch times reveals that the transport took 10 minutes longer than is normal to reach the destination hospital.

1. **If you do nothing, what will happen?**
2. **What do you say or do?**
3. **With whom do you address the issue?**
4. **What are two other alternatives you could use?**
5. **What is your last-resort action?**
6. **If you reach your last-resort action, what effect will you be dealing with two weeks after taking action?**

20. Disappearing Station Supplies

▸ NFPA 1021: 4.2.6, 5.2.1

At one of your fire stations, supplies seem to be disappearing. Because station supply order forms are submitted to you for processing and delivery of requested station supplies when you make rounds, you are aware that certain common supplies are being ordered almost weekly for a station that should have an ample supply. Requisitions for toilet paper, kitchen scrub pads, dish soap, shop rags, vehicle wax, and even copy paper now seem to be submitted every 9 to 12 days. This was not always the case; indeed, the requests have become noticeably more frequent within the past three months or so. At first, you dismissed the unusual volume of requests as an anomaly. When you decided these requests must be a carry-over from other shifts and considered denying the supply requisitions as duplicates, your company officer, Sergeant Lucas, told you that she does not know where things are going. It is essential for the members of your shift to have the necessary supplies to complete their assignments, and the sergeant needs your help.

1. **If you do nothing, what will happen?**
2. **What do you say or do?**
3. **With whom do you address the issue?**
4. **What are two other alternatives you could use?**
5. **What is your last-resort action?**
6. **If you reach your last-resort action, what effect will you be dealing with two weeks after taking action?**

21. Lieutenant Under the Influence: Not Formally Reported

▸ NFPA 1021: 4.2.4, 4.2.5, 5.2.1

You receive word through informal channels that one of your station officers is regularly noticed "popping pills" while on duty and at times acts "a little foggy." This information has come up in casual conversation around the station while you make your daily rounds, but no one has directly told you who, when, what, or how often. You have repeatedly overheard your crew refer to an "Officer Foggy Lou," and today you heard a fire fighter say, "If anything's ailing you, the guy to see is the lieutenant. He has a prescription for everything."

1. **If you do nothing, what will happen?**
2. **What do you say or do?**
3. **With whom do you address the issue?**
4. **What are two other alternatives you could use?**
5. **What is your last-resort action?**
6. **If you reach your last-resort action, what effect will you be dealing with two weeks after taking action?**

Attitude Issues

22. Disgruntled Company Officer

▸ NFPA 1021: 5.2.1

Two weeks ago, Lieutenant Ash was moved to your shift from another shift. He has a reputation among his former co-workers as a complainer, and even some of your fire fighters are aware of his constant complaining because they have previously worked with him on overtime or through shift trades with other fire fighters. Your shift personnel have also previously heard Lieutenant Ash grumbling at shift change. He has a long-standing reputation for continually making comments around the station about his dislike for the department and departmental decisions and direction. Targets of his discontent include the people he is assigned to supervise; the equipment the department purchases; the design of the apparatus, from water tank size to hose capacity; and even the things that wear out around the station and need repair. Lieutenant Ash does not hesitate to express his displeasure to his crew or his comrades working on the shifts before and after his own shift.

You are now Lieutenant Ash's supervising battalion chief, and he continues to gripe to those around him in his new assignment. His fire fighters have informed you during your regular station visits that this officer is not a happy man, and his negativity is going to be a challenge to them. Even the personnel from Lieutenant Ash's old shift, when you encounter them, mention that they are glad he is no longer a part of their shift, because he brings everyone's attitude down. You are unaware of any previous actions or discussions with Lieutenant Ash instigated by his former commanding officers about his behavior. You recognize his behavior can be a drag on morale on your shift, and it is evident from fire fighters' comments that department morale has already been affected.

1. **If you do nothing, what will happen?**
2. **What do you say or do?**
3. **With whom do you address the issue?**
4. **What are two other alternatives you could use?**
5. **What is your last-resort action?**
6. **If you reach your last-resort action, what effect will you be dealing with two weeks after taking action?**

23. Highly Qualified Company Officer Passes up Promotion Opportunities

▸ NFPA 1021: 5.2.3, 6.2.4

Captain Salitto, a company officer in your department, handles almost every assigned task or issue well. Nevertheless, when promotion opportunities arise for battalion chief or a staff officer position, Captain Salitto consistently chooses not to seek promotion, despite encouragement from other officers and fire fighters. As a result, the candidates who are promoted within the department, although qualified, are all significantly less qualified than Captain Salitto.

1. **If you do nothing, what will happen?**
2. **What do you say or do?**
3. **With whom do you address the issue?**
4. **What are two other alternatives you could use?**
5. **What is your last-resort action?**
6. **If you reach your last-resort action, what effect will you be dealing with two weeks after taking action?**

24. Budget Cuts Reduce Staffing and Morale

▸ NFPA 1021: 4.2.6, 5.2.1, 5.4.6, 6.2.1, 6.4.6

Budget cuts have reduced your company staffing from four crew members to three, and your entire shift is upset. Several of the companies' members had worked together for many years, during which time they developed into effective response companies and became close friends. The remaining members are upset about having lost their colleague. Judging from their expressions, kitchen table chatter, and lack of enthusiasm for their tasks, you know that some members of the companies are mad and discouraged. They are also concerned that with the reduced staffing, company assignments to complete tasks on emergency scenes will take longer, and they fear that they may even risk injury while trying to do so much in the first few critical minutes of the response. Sometimes, their discouragement boils over into anger at the circumstances that brought about the reduction in staffing. The staffing reductions were enacted only a month ago, so their full effects are only now becoming evident. You knew the staffing reductions were unpopular with the entire department even before they were implemented. The issue was bigger than just your shift and your shift management responsibilities, so you had not previously taken any actions. Now that morale and the personnel resources needed to conduct your shifts' activities are affected, however, you have a shift issue to manage as the shift's leader.

FIGURE 3-9

1. **If you do nothing, what will happen?**
2. **What do you say or do?**
3. **With whom do you address the issue?**

4. **What are two other alternatives you could use?**
5. **What is your last-resort action?**
6. **If you reach your last-resort action, what effect will you be dealing with two weeks after taking action?**

25. Company Dissatisfied with Design of New Apparatus

▸ NFPA 1021: 5.2.1, 5.4.6

A new piece of fire apparatus is assigned to one of your stations, Station 14, and your Engine 14 crew members do not like the newly received fire engine's compartment design. They think the storage compartment located inside the cab is both too big, reducing the room available for the crew, and poorly designed, because the equipment intended to go inside that compartment could just as easily be stored in other places. You are aware that Engine 14's crew is unhappy with the apparatus design: As you have visited the other stations under your command, you have heard fire fighters who have occasionally been detailed to Station 14 to cover an absence talk about the complaints Engine 14's crew have made to the Engine 14 lieutenant. This is the first brand-new apparatus the station has received in years; Station 14 has historically received apparatus transferred in from one of the busier stations within the department when those stations received a new apparatus.

1. **If you do nothing, what will happen?**
2. **What do you say or do?**
3. **With whom do you address the issue?**
4. **What are two other alternatives you could use?**
5. **What is your last-resort action?**
6. **If you reach your last-resort action, what effect will you be dealing with two weeks after taking action?**

26. Inappropriate Comments About a Patient During a Response

▸ NFPA 1021: 4.6.3, 5.2.1

You respond as the command officer to a motor vehicle collision with possible entrapment. Engine 2, Rescue 7, Ambulance 2, Ambulance 4, and law enforcement personnel are on scene when you arrive. The female driver of one of the vehicles is still inside the vehicle and is noticeably impaired. She smells strongly of alcohol, and empty beer bottles are visible in the vehicle. Two female passengers from the same vehicle are being packaged for transport with serious but non-life-threatening injuries, including trauma to extremities and lacerations. These passengers also smell of alcohol and are crying excessively. As the driver is being extricated, you hear Dave, one of the members from Rescue 7, comment to Mary, a fire fighter from Engine 2, "It's gals like this that keep us in business." There are spectators around the crash scene, and you are unsure if anyone has heard the comment. Nevertheless, it was spoken loudly enough for you to hear it, and some of the other fire fighters working around the car look up at Dave in response to his comment. Mary just makes eye contact with Dave, does not comment, and keeps working.

1. **If you do nothing, what will happen?**
2. **What do you say or do?**
3. **With whom do you address the issue?**
4. **What are two other alternatives you could use?**
5. **What is your last-resort action?**
6. **If you reach your last-resort action, what effect will you be dealing with two weeks after taking action?**

CHAPTER 4

Battalion Officer/ Shift Commander Discussion

Work Product Issues

1. Dirty Dishes

▸ NFPA 1021: 4.2.2, 4.2.4, 4.2.6, 5.2.1

Your department expects that every individual will clean up after himself or herself. Chore lists, supervision, and history all reinforce the assumption that a fire station's kitchen, along with the rest of the station, is to be kept clean and orderly. You receive a complaint from a company officer, Lieutenant Malone, that the offgoing shift routinely leaves a mess in the station kitchen. The oncoming crew arrives to find the floors unswept, the table dirty, and dirty dishes left in the sink. A discussion between your company officer, Lieutenant Malone, and the offgoing company officer, Lieutenant Redding, has not resolved the issue.

1. If you do nothing, what will happen?

If you do nothing, your subordinate officer, Lieutenant Malone, may lose confidence in you as his supervisor. You also risk losing the confidence of your operating personnel because it is highly likely that the fire fighters know their officer has tried to correct the situation with no success and has come to you for help. Finally, if you do nothing, you risk increasing dissension among shifts in the department because Lieutenant Malone and Lieutenant Redding have been unable to find a common ground to resolve the issue. Other than leaving the attitude and cooperation concerns unaddressed, if you do nothing, the oncoming crews will probably eventually clean the kitchen to some acceptable level. You will not likely see any loss of community life or property over the matter.

2. What do you say or do?

You must recognize that the issue has been brought to you for assistance. It is important for you to help but not necessary for you to solve the problem or fix the issue yourself. This small-scale station issue provides you, as the shift commander, with an opportunity to educate your officer(s) that these types of in-house, non-life-threatening issues should be addressed at the company officer level. You may help by communicating clear expectations to Lieutenant Malone and simultaneously communicate with your peer, the battalion commander whom you are relieving, that Lieutenant Malone is having difficulty resolving an issue with that commander's company officer, Lieutenant Redding. This discussion may be followed with a request from you for the other shift commander to follow up with Lieutenant Redding to assess what exactly is occurring. By engaging the peer shift commander, you have assisted Lieutenant Malone without stepping in and handling the issue for him.

3. With whom do you address the issue?

First, communicate with Lieutenant Malone that his concern has been heard. In this discussion, emphasize the importance of a company officer's ability to solve problems and resolve issues. Offer guidance to Lieutenant Malone about which steps he might take to continue to work through the issue on a peer-to-peer basis with Lieutenant Redding. This discussion should be held with Lieutenant Malone alone, and it offers an opportunity for you to coach your company officer on how to push through obstacles or lack of cooperation. It should not involve other members of the company; broadening the involvement could create hostility toward others on the other shift or reveal to the company any shortcomings of the company officer if every alternative has not yet been attempted. Coach your officer privately.

You should also provide support to Lieutenant Malone by mentioning the situation to your peer shift commander on the preceding shift. Inform him or her that Lieutenant Malone has reported some unresolved station conflict but that he is continuing to work on a resolution. You could ask the shift commander to assess how things are going at that particular station and whether his or her company officer, Lieutenant Redding, has any unresolved issues. It is important to communicate to the peer shift commander that this is neither a request for intervention on his or her part nor a complaint; rather, your notification is simply meant to keep both of you informed and aware.

4. What are two other alternatives you could use?

One alternative is to make an unannounced friendly visit to the station just before shift change to assess the situation yourself. Your visit should not be preannounced to either your shift or the preceding shift, nor should any of the crew members know what you are looking for. You may discover that the alleged mess is nothing more than a few coffee cups in the sink and a stray crumb or two on the counter or stove. To some people, anything less than perfect constitutes "a mess." If your company officer or crew member has extremely high expectations, the reported mess may be nearly impossible to completely correct. This does not mean Lieutenant Malone is in the wrong, but it may explain why Lieutenant Redding does not seem to believe that there is an issue to be addressed.

A second alternative is to instruct Lieutenant Malone to document the kitchen messes, with dates and details, and possibly including photos, and then approach Lieutenant Redding with the evidence. Lieutenant Redding may have discounted the issue as not as frequent or severe, in which case anecdotal evidence can provide a good reality check. In fact, such documentation may reveal to Lieutenant Malone that the problem is not as prevalent as it seems. If you encourage your officer to choose this option, it is important for Lieutenant Malone to not draft a memo, corrective action form, or letter of interrogation; rather, a simple handwritten note on a notepad or an informal digital note is sufficient. Remember—this is a station issue that really needs to be solved at the station level.

5. What is your last-resort action?

Should Lieutenant Malone attempt numerous alternatives without resolution, and you have found the issue to be a legitimate source of dissension, you might consider approaching your peer shift commander on the preceding shift and requesting that he or she become engaged in the issue. Such a request should emphasize the importance of having the two company officers, Lieutenant Malone and Lieutenant Redding, work things out between themselves, but note that you would appreciate some pressure on Lieutenant Redding to cooperate in the dialogue and problem-solving process. You are not requesting for your peer shift commander to get involved in solving the officers' issue, as this particular issue must be resolved at the company level. It is important to have the company officers keep working toward a resolution, because this effort will help them develop their own problem-solving skills, and any action they take will have very little detrimental effect. In contrast, if a shift commander intervenes and resolves the situation for a company officer, the company officer may start to develop a pass-it-off problem-solving style. The pass-it-off attitude is not helpful for the officer, who should learn to resolve issues on his or her own, nor is it productive for the officer receiving the hand-off, because it creates significantly more work at a higher level that should be handled closer to the issue at hand. A messy kitchen is not a shift or battalion commander–level issue.

6. If you reach your last-resort action, what effect will you be dealing with two weeks after taking action?

Depending on the discretion of your peer shift commander, he or she may quietly address the issue with Lieutenant Redding, and the two company-level officers may have either resolved, or be in the process of resolving, the kitchen issue. However, once the other shift is engaged from the shift commander level, tensions may develop between the shifts. Lieutenant Malone's shift may be perceived as having unnecessarily complained to their shift commander and having reported station issues outside the station. Much will depend on the company officers' tact and professionalism in this situation. Good company officers will work to resolve issues without pitting crew against crew or shift against shift. Such discretion is rare, however; camaraderie among the crews, including the company officer, generally outweighs officers' camaraderie and willingness to cooperate and work through the situation. This issue is a great opportunity to educate your company officers about the value of developing a stronger sense of camaraderie with their peer officers than with the crew they supervise.

2. Low Fuel Gauge

▸ NFPA 1021: 4.2.2, 4.2.4, 4.2.6, 5.2.1

A company officer, Captain Garcia, reports to you that the pumper is not being fueled regularly, and his shift is consistently noting a low fuel gauge during their morning apparatus checks. Captain Garcia has discussed this issue with the offgoing company officer, Captain Smith, but the situation has not been corrected.

1. If you do nothing, what will happen?

Although it is not likely the low fuel level will prevent the apparatus from making responses, the station's crew members and the company officer, Captain Garcia, will grow increasingly frustrated if the situation continues. This creates additional work for the company if they are going for fuel more often; however, it is not likely to take them out of response mode. This issue affords you, as the battalion chief, the opportunity to develop your company officer. This is a common station-based issue that calls for the company officer to work his or her way through the challenge.

2. What do you say or do?

First, ask Captain Garcia some direct questions: How often is the apparatus being left low on fuel? How low is the fuel gauge when you find it this way? Then shift the focus by asking, "What has the preceding shift's company officer, Captain Smith, told you about *why* the apparatus was low on fuel?" The typical response is likely to be something along the lines of "She didn't say anything about the apparatus being low on fuel; we just keep finding it that way." Instruct your officer to ask the preceding shift's company officer about the situation—specifically, *why* the apparatus is low on fuel—and give her a chance to explain. Sometimes fire fighters and officers assume that because an issue has been referred to in passing—"We found the apparatus low on fuel last shift when we relieved you" or "I wonder if we will have to go get fuel in the apparatus . . . again"—the issue has been communicated and will be fixed. On the other side of the issue, you may find that the offgoing shift may have been up three or four times throughout the night and assumed the oncoming shift would be willing to help out by fueling up when they start the shift. They may have communicated this assumption to the oncoming shift throughout indirect comments ("Had a busy night last night—four calls after midnight") without specific mention of the need for fuel. An assumption such as this is a coaching moment for you and your company officer.

3. With whom do you address the issue?

Speak with Captain Garcia alone. Talk through the issue of assumptions versus direct questions, and assess the severity of the fuel level problem. If the apparatus is left on fumes and the ability to make a response is jeopardized, you may have a different situation, but rarely are station-related issues so extreme. Many departments have regulations that require refueling when the apparatus is three-fourths or one-half full. As battalion chief, you need to ensure that the conversation remains between you and Captain Garcia. Allow him to solve the issue. If your department does have a policy specifying when apparatus is to be refueled, make that document available to Captain Garcia for him to discuss with Captain Smith.

4. What are two other alternatives you could use?

One alternative would be to reinforce expectations. When the apparatus is found fueled up and ready, consider acknowledging that state by saying "thank you" or dropping off a snack for the preceding shift to enjoy. Although these gestures may seem a little excessive for completing a task that is part of the preceding shift's job, the question for Captain Garcia is whether he wants to change the shift's behavior or fuel the apparatus. If the oncoming shift acknowledges the offgoing shift for their contribution, they may just make a point to leave the apparatus in great shape for your shift. Captain Garcia might also consider sending an e-mail to Captain Smith's battalion chief praising Captain Smith's efforts to leave the apparatus or station in great shape for the oncoming shift. Sometimes a simple thank you goes a long way.

A second alternative might be to have Captain Garcia discuss the issue with the company officer who comes in after your shift. See if the two officers can reach an agreement to focus on the fuel levels so as to ensure the apparatus is full and ready to go for the oncoming shift. Sometimes members of a shift feel that if an apparatus was left for them with low fuel, then they will leave it in the same condition for the ongoing shift. This is not the best way to handle things, but it is more common than anyone wants to admit. If the apparatus is short on fuel regularly, that problem may stem from the way Captain Smith finds the apparatus when she comes in. Is it possible that it was low on fuel for Captain Smith's shift but those shift members were never called out of the fire station, so they did not make it a point to refuel? Perhaps the culprit is not the preceding shift at all. Maybe a three- or four-way pact involving all the shifts might resolve the situation. Encourage Captain Garcia to think outside the issue. He might even start a friendly competition to see which shift can put the most gallons of fuel in the apparatus during their shift for the week or month. In the best-case scenario, you might end up with crews conducting driver training just to try to get more gallons in the tank.

5. What is your last-resort action?

The last-resort action is for you to get involved as the battalion chief. If you become engaged in the issue, it means the company officers have been unable to resolve it on their own. This is a concern for an issue that affects response readiness, shift-to-shift cooperation, and company-level leadership problem solving. Should you be required to intervene, the first step might be to stop by the station a few shift changes in a row and assess the situation for yourself. You can do so quietly, without drawing the attention of either your shift or the preceding one. Investigate where the root issue lies so you can address it. If you find that Captain Garcia has overstated the problem, counsel him to use a reasonable standard and not a self-imposed, more restrictive one. If the apparatus is being left low on fuel, consider having Captain Garcia ask Captain Smith to explain the situation in your presence. If you decide to take this course of action, make sure Captain Garcia knows he will lead the meeting and that, as the battalion chief, you are simply there to listen. Again, allow the weight of the position to communicate expected cooperation, but try to keep as much problem-solving power at the company officer level as possible.

6. If you reach your last-resort action, what effect will you be dealing with two weeks after taking action?

You may see some nit-picking among shifts. This kind of small-minded response is not unusual, and you can use this opportunity to reinforce with Captain Garcia the importance of not becoming part of the pettiness. Encourage Captain Garcia to look for the right things to do: clean what is dirty, fill what is empty, fix what is broken and can be fixed, and make sure that the company is completely response ready for the public when Captain Garcia and the company may be called upon—and do those activities regardless of who brought them up or why.

3. Changes to Run Card Assignments

▸ NFPA 1021: 4.2.6, 5.2.1, 5.4.6, 6.2.1

One of your company officers, Lieutenant Vickers, constantly offers ideas to change the running card assignments used by the computer-aided dispatch system on the grounds of efficiency. It is not unusual to receive an occasional suggestion from a company officer for a change to a run card, but most previous suggestions have been based on new roads or road construction. These infrequent changes were relatively obvious and were consequently implemented. The changes that Lieutenant Vickers is suggesting, however, consistently create a decreased call volume for his company.

1. If you do nothing, what will happen?

If you do nothing, your inaction may maintain the status quo operationally, but it will increase the risk of developing a frustrated company officer. Remember, Lieutenant Vickers is positioning the suggestions as ways to improve response efficiency. The fact that these changes would reduce the number of his company's potential responses may appear self-serving, but be careful not to jump to conclusions. Even if no one else on Lieutenant Vickers's crew or the lieutenant's peers believe that run card changes are merited, Lieutenant Vickers's frequent suggestions for changes will not decrease if you fail to respond to them. In fact, if you do not act on his suggestions, you may create frustration in that Lieutenant Vickers may feel slighted—that is, he might perceive that his suggestions do not seem to merit any action from his supervisor.

2. What do you say or do?

First, you must assess the merits of the suggested changes. If you determine that they will not have benefits and do not communicate your assessment to Lieutenant Vickers, he will continue to make suggestions. Conversely, if you determine there are no benefits other than the self-serving ones for Lieutenant Vickers's company, and this assessment causes you to miss a true opportunity to increase response efficiencies, then you as the shift officer have now contributed to the inefficiency of the department. You should ask Lieutenant Vickers to fully explain the reasoning behind his suggestions. Inform him that he must thoroughly research any suggestion before bringing it to you. When you or the other supervising officers who may be consulted decide whether to change run card assignments, it will be with full knowledge that the lieutenant's ideas and suggestions were considered by reviewing all the facts. Explain to Lieutenant Vickers that, after consideration of his initial input, once a decision is made, the matter will not be open for further suggestion unless a variable changes, such as opening a new fire station, relocating a fire station, or a change in response personnel levels.

3. With whom do you address the issue?

Speak directly with Lieutenant Vickers. The company officer is in the position of leadership for a reason, and such projects and communications should initially be between the battalion officer and the company officer only. If the suggestions merit review by peer battalion officers or the battalion officer's supervisors, you should communicate back to Lieutenant Vickers that such discussions are progressing through the chain of command.

4. What are two other alternatives you could use?

One alternative is to harness Lieutenant Vickers's enthusiasm to look holistically at the entire jurisdiction to consider where run card changes might benefit the community served by the department—for example, where run card assignments could be bolstered by adding them to busy districts. This effort would entail gathering data to show cause-and-effect relationships for all companies. The assessment might also involve considering whether Lieutenant Vickers's station is still needed in the present location. This alternative could create significant work for others, including information technology staff, data management personnel, and even other fire officers, and should be carefully considered before pursuing it. If Lieutenant Vickers's suggestions have merit, however, you have an obligation to consider possible alternatives that could provide better service to the community, regardless of how the individual company fares with call volume.

A second alternative is to provide a definitive answer to the constant suggestions. Communications are often misunderstood. You might think that a casual comment to your officer about how his company will have fewer calls with the suggested changes and a jest that you are "on to his plan" would be enough to have him drop the idea, but you have not really provided your officer with an answer but rather just a sarcastic remark. While you may intend for such indirect comments to be perceived as an answer, they are likely to leave your officer waiting for or communicating other suggested changes until you provide a definitive answer. A high-level officer may provide feedback such as "Let me think about it" or "I'll look into it," intending that response to mean that the issue is resolved and no changes are needed. The lower-level officer, however, might assume that you will take action at some point, and he or she may continue to offer additional suggestions. To avoid this kind of misperception, be direct with your answers, and, as the old saying goes, say what you mean and mean what you say.

5. What is your last-resort action?

The last-resort action may be contingent upon the merits of the suggestion. If the response changes are plausible, you might consider seeking counsel from a peer shift commander to see if the suggestion would be workable across the board. This is a later-arising alternative, because to reach it, you should have already concluded that Lieutenant Vickers's idea is worthy of bringing more minds to the matter. If it is greeted with interest, you may start the process of requesting policy changes in accordance with your department procedures. Conversely, if the suggestions have little or no positive impact on responsiveness to the community and benefit only Lieutenant Vickers's company, your last-resort action might be to have a private meeting with Lieutenant Vickers and brief him on your decision. Be sure to thank him for his willingness to provide you with suggestions. Explain directly how you reached your decision, using appropriate facts such as response data, response times, secondary response data, and other factors that you considered in your decision-making process. Then be direct and explain that this issue is closed for discussion and there will be no further consideration given to changes, which means there is no need to continue to offer suggestions for this particular issue unless a factual variable changes. If such a variable should change, ask Lieutenant Vickers to brief you on the change; you as the shift commander can then determine if any response adjustments are necessary. This process allows your officer to understand you are not unwilling to listen but you are unwilling to revisit the same issue repeatedly.

The challenge with this alternative is to be firm enough to convey the finality of the decision yet diplomatic enough to preserve the working relationship with this company officer so that Lieutenant Vickers feels his input is valued. This balance has the ability to impact decisions and suggestions in all aspects of the job in the future. You do not want to quash Lieutenant Vickers's spirit, yet you do not want to provide false hope or string him along.

6. If you reach your last-resort action, what effect will you be dealing with two weeks after taking action?

The result two weeks later will be determined to a great extent by the amount and sincerity of the communication you have with your company officer during the one-on-one meeting. If you are direct yet open and empathetic, you will likely have a positive outcome: The response running assignments will be unchanged, Lieutenant Vickers will no longer constantly suggest different running assignments, and Lieutenant Vickers will perceive that, despite the rejection of this particular suggestion, you are open to considering other future suggestions.

In contrast, if you appear frustrated by the response change requests or suggest they are self-serving for Lieutenant Vickers or if the one-on-one meeting is interrupted by calls or other appointments, the company officer may feel resentful. The inability to communicate and completely address the issue at one time will likely create a negative atmosphere and the perception that you do not welcome suggestions. If the conversation is frustrated or interrupted, in two weeks, there will likely be anxiety or possibly anger brewing within the company and looking for a place to spread. It is imperative that you arrange to have enough time to work through this issue completely with your company officer. This task and responsibility reside with you as the shift commander, not the company officer.

4. Negative Feedback from Personnel

▸ NFPA 1021: 4.2.4, 4.4.2, 5.4.5, 6.4.6

Chief Willis, the operations chief, assigns you the task of assessing morale within the department and reporting your findings to him with suggestions for improvement. As you begin to make inquiries among the companies, you sense some reluctance by the line personnel to answer your questions. You decide to supplement your interviews with an anonymous survey instrument regarding the climate of the department. You are shocked to receive an overwhelming number of responses from personnel suggesting that you are the source of the department's problems. Some believe that you are inconsistent in how you treat people and that you are selective regarding accountability. Some believe your supervision is unfair. Others report they do not trust you. The line personnel seem to have a perception that you are not proactive when it comes to department needs, such as repairs.

1. If you do nothing, what will happen?

It is important to recognize that if you do nothing, your inaction will create two significant problems. The first and most direct problem would be a failure to carry out a legitimate assignment given to you by Chief Willis. Even if the results from the questions you asked verbally and the results received from the supplemental anonymous survey questionnaire are not what you wanted to hear, you must complete your assigned task. The second problem if you were to do nothing would be the erosion of the department's effectiveness. Keep in mind that you have received feedback from the personnel identifying a perceived source of a problem. By doing nothing, you convey the message that feedback or input from personnel will be accepted only if it is favorable. If you do not act, that lack of response devalues the personnel and allows the attitudinal climate to decline rapidly.

2. What do you say or do?

First, you must take a moment to prepare yourself for the next steps. Being identified as part of a problem is not pleasant. It is a natural human reaction to want to defend your actions or inactions, dismiss the validity of the views voiced by the respondents, or even lash out at those personnel who share these views with you. It is extremely important that you recognize these reactions as normal human responses—but you must also press on with the courage and objectivity of a fire officer. Just as dealing with injury or death at an emergency scene requires a conscious effort to focus on the objective task to complete the mission, so you must apply the same courage and objectivity to this station-based issue, and then take the next step.

That next step should include digging deeper into the responses. You will need to ask follow-up questions. If the responses you received were related to personnel or issues beyond your immediate control, you should gather information on those issues, report it, and have appropriate personnel act within their capacity. In this situation, however, you have been identified as a source of the issues, which include inconsistency, untrustworthiness, unfairness, and lack of accountability. Consequently, you have the ability to effect changes directly.

Ask for detailed feedback in person if you are able to set aside some time with the personnel providing these responses. If the anonymous survey yielded these results, you might have to meet with all your crews and advise them of the results and verbally ask follow-up questions seeking further input: "Can you provide specific examples that support these statements?" or "Are these examples or observations your own, or has this information come from other sources, such as peers, other companies, or shifts?" Remember the task assigned to you by Chief Willis included not only an assessment of the department's morale but also the development of suggestions based on your findings. Since you have been identified as at least part of the problems by some respondents, you can capitalize on this opportunity to investigate possible suggestions for improvement by the field personnel. You might ask, "Which one or two things do you feel might improve the situation?" or "If I were to make changes to improve the situation, which change should I make first?" The answers to these questions will provide valuable information to help you complete your task of suggesting solutions to the climate-related problem. You might not choose to report every small detail you receive, especially if you receive conflicting feedback, but you will have to analyze that feedback and report the significant issues or recurring themes identified from your initial and follow-up questions. You may want to implement certain changes immediately, but you may also identify some perceived issues that are not based on possession of a complete set of facts. You will have to determine how to best utilize the results of your inquiries.

After gathering the additional information, it is very important to let the personnel know you will share this information with Chief Willis. This message conveys two important facts. First, the operations chief is clearly concerned with the climate of the shift or department. This is a positive organizational dynamic and should be leveraged for the good of the department. Second, you will complete the task assigned to you effectively, regardless of the personal gain or loss associated with its outcome. This fact can lead to an instant positive change by communicating to your personnel that you place the good of the department above your own interests. Keep in mind that by communicating with your personnel, you are not committing to any changes, but rather you are reporting the results of your findings from the field, as assigned.

3. With whom do you address the issue?

Communicate the questions and responses identified through your follow-up efforts to everyone whom you originally polled to assess the department's climate and morale. Ideally, you should ask follow-up questions at the time you receive the initial feedback or a few days to a week later but no longer than two weeks from the initial inquiry. Keep the process timely and relevant to ensure you do not become associated with ineffectiveness.

You might also consider asking some people on other shifts to see whether the negative responses are isolated to a particular shift. Assessing all shifts may or may not be part of your original assignment from Chief Willis, but if you have received the assignment for your shift only, the additional shifts' perspectives may provide some insight. If you ask the other shifts for input, it would be appropriate to talk with your peer shift commander first to allow him or her to understand you are gathering information on yourself and not trying to usurp his or her assignment or authority.

Last, communicate the results to Chief Willis. Complete the task with a report of findings and suggestions, as originally assigned.

4. What are two other alternatives you could use?

One alternative might be to take the raw results you found, with very little additional probing of your personnel, and include them in the report, along with your suggestions of how to address the situation. Although personnel identified you as a significant part of the problem of low morale due to your inconsistency, untrustworthiness, unfairness, and failure to assign accountability across the board, you may believe their responses have limited merit. For example, you might have additional information—of which the field personnel are unaware—that you use when making decisions. Thus, you may decide the issue is one of perception rather than an actual need for changes. To maintain the integrity of the process, however, you must be sure to include all the results of your assessment, both favorable and unfavorable. Suggestions might include providing additional communication to the field personnel so that their perceptions of you and your actions will not be based on incomplete, inaccurate, or nonexistent facts. This alternative completes the task assignment but will do little to impact your shift in the short term. It also leaves some risk unaddressed, since your supervisor, Chief Willis, will have to assess the validity of the results, including your role, on his own.

A second alternative might involve digging beyond the initial responses to gather additional information. For example, if your initial inquiry results in responses that personnel perceive you to be "unfair" because you do not discipline people "fairly," you might dig below the surface to assess which issues you are acting on with corrective action and how many other instances you are not acting on, possibly because you are unaware they even exist. Is there an attendance issue, a reporting issue, or a punctuality issue that might be contributing to discontent rather than your actions or inactions? For example, perhaps fire fighters at one station are running late more often than their peers at other stations, and line fire fighters think you are aware of this problem but ignoring it, when you really were not even aware these issues were occurring at all. In contrast, when you learned of tardiness in a different station, you might have taken prompt action through your station officer—a response that propagates the perception that you are inconsistent and unfair with enforcement of regulations. If you use this alternative, you must be careful not to fall victim to your natural human instinct to become defensive. If you look for underlying issues, be sure your quest is not really an attempt to cover up or make excuses for issues associated with you.

5. What is your last-resort action?

Your last-resort action might include reflecting on your abilities and performance in the context of the organization. Are you in step with the organization and its mission? Has the department's mission changed, yet you have not changed with it? Can you change? If you are unknowingly contributing to the problems, and the feedback you have received is determined to be legitimate, are you willing to take steps to modify your methods and behavior to create a positive and more effective workplace? If you are unable or unwilling to make modifications, you might consider whether it is time to make a change for yourself. Such a change could include a positional change that involves fewer responsibilities with crew or shift supervision; consideration of different responsibilities such as a transfer request to a different division; or, in extreme situations, retirement. These options are dramatic steps that are not to be taken lightly. You must assess how severe the problems are and consider solutions appropriate to the level of issue.

If true inconsistencies in your behavior have resulted in loss of trust with the membership, it is not too late to make changes if you are willing. Change is not easy, especially when dealing with behavior. If the change is too much for you to embrace and accomplish, it is important to acknowledge as a last resort that to improve the situation, you may need to consider reassignment. Honest dialogue with your supervisor about the results of your assignment may yield more insights as to how you could make adjustments that positively affect the entire organization, including you.

6. If you reach your last-resort action, what effect will you be dealing with two weeks after taking action?

You may be dealing with dramatic effects if you end up considering retirement or position reassignment. These changes could impact your schedule, your pay, and many other people, such as your family and co-workers. Reassignment could impact the department in terms of promotional opportunities and possibly even more challenges as the organization deals with the uncertainties of significant change. You may also experience some "back-pedaling" if personnel who originally identified you as part of the problem start to regret the consequences of their criticism. Such back-pedaling can be frustrating for you or even cause you to become angry. Make sure the information you use to make your decisions is factual rather than negotiable. It is equally important that you are comfortable with the decisions you make before taking any dramatic last-resort actions.

5. Rearranged Apparatus Equipment

▸ NFPA 1021: 4.2.6, 5.2.1, 5.4.6

A new ladder truck has been assigned to Station 3, and your shift's crew members at Station 3 do not like the location and arrangement of equipment within the apparatus. You were aware from making your routine station rounds that the crew of Station 3 was not in complete agreement with some of the equipment placements on the ladder truck, but this is the first time you have heard that the crew has been rearranging equipment for use during its shift. The battalion chief from the shift following yours, Battalion Chief Striker, informs you that his company officer at Station 3 has reported that his shift regularly arrives to find that the equipment on Ladder 3 has been rearranged and is not in the locations where it is supposed to be stored on the apparatus. The issue comes up daily, and the crew has strongly expressed their anger to their company officer.

1. If you do nothing, what will happen?

If you do nothing, the effective relationship between the shifts will deteriorate. This particular issue affects all members assigned to that station but most critically those on Ladder 3. Legitimate authority exists for the station officers and crew members to carry out their assignments with the equipment and apparatus provided. If you do nothing, you are communicating that you are not willing to attempt to solve problems, and that message also devalues the credibility of your counterparts on the other shifts. Although this may not be your intention, when a fellow shift commander approaches you to discuss a situation, you must realize this issue is serious enough for a crew member to have taken it to his or her shift commander and for your peer shift commander to subsequently have brought it to you. If you do not act, you communicate a very strong message of complacency.

2. What do you say or do?

Just as when important details are conveyed to you on an emergency scene, you should first acknowledge the message was received. Confirm with Battalion Chief Striker that you received and understood his message and will give it your attention. Second, you should thank your fellow shift commander for coming to you with the issue. Battalion Chief Striker brought you bad news rather than avoiding bringing it to your attention, which could have allowed the issue to fester without your knowledge. He values your input and believes you will be committed to solving the issue. Rather than taking the matter up the chain of command or using his influence as a higher ranking officer to deal with your station officer directly, the battalion chief brought the matter directly to you. You can enhance this rapport by simply acknowledging your appreciation of his actions and expressing your willingness to engage in evaluating the situation and working toward an agreeable solution. You do not need to have answers at this stage of the conversation; you will still need to research the issue.

The next step is to inform your shift officer at Station 3 that you have been informed about some disagreement with the shift's equipment placement on the new ladder truck. You can inform him without accusing or dismissing the issue, but delivering this message allows you to begin candid conversations so you can gather facts and assess the situation for yourself.

3. With whom do you address the issue?

Acknowledge the message and express your appreciation directly to your fellow shift commander. You can do so during the same meeting in which he brings you the news. Also consider conducting a follow-up exchange in a private setting at the next shift change, when you can reiterate your appreciation that he has brought the issue to you. Rather than rely on a passing comment you made at the close of your initial conversation, an explicit follow-up further conveys your sincerity and can help to increase your rapport with your fellow shift commander.

Next, you should communicate with your shift officer in Station 3. This conversation should be held directly between you and the station officer only. Sometimes it may seem more efficient for the shift commander to go directly to the apparatus operator or fire fighter who seems to be driving the issue for firsthand information. In reality, such a breach of the chain of command erodes the shift commander's time management, diminishes the company officer's effectiveness, and can send an incorrect message to the crew that they have influence over the situation because the shift commander is dealing directly with them. Give the station officer the same consideration the battalion chief gave you by going directly to him.

4. What are two other alternatives you could use?

The first alternative might be to research any applicable memoranda, standard operating procedures or guidelines, or similar documents that might indicate any clear-cut direction that is established policy and not open for interpretation. If any such references exist, you could assess the equipment placement for yourself while the crew is in training or on a detail so you do not raise concerns or questions prematurely. A private investigation of the equipment placement will allow you to gain insight into the facts without inciting concern among personnel. Fire fighters watching a shift commander open and close compartment doors on an apparatus will question the reason behind such activity and may jump to dramatic or punitive assumptions. If you find equipment is out of place, do not jump to conclusions before hearing the "who, what, when, where, and why" from your company officer.

If you find equipment has been moved out of its assigned position on the apparatus and clear directives exist regarding the appropriate location of this equipment, then the placement must be corrected. This step does not address the intent or validity of a requested change, but to ensure the effectiveness of the entire organization, all equipment must be placed in its assigned locations until reference documents can be reviewed and changes considered and implemented if appropriate.

A second alternative would be to engage in a direct conversation with your company officer, informing him that you have been made aware of equipment placement issues that seem to be creating disagreement. Assign your company officer to sketch out the apparatus and, along with his crew, identify in writing every piece of equipment to be placed on it according to currently assigned compartment and cab locations. Then have the company officer identify in writing his proposed changes, along with the rationale for those changes.

Finally, instruct the company officer that until the project is complete and any proposed changes submitted for review and subsequently approved or denied, the equipment on the apparatus will remain in the currently assigned locations. This alternative allows you, as the shift commander, to ensure consistency among response forces, maintain effective relationships from shift to shift, and still leave opportunities open for the companies' ideas that could yield operational improvements. Your action also communicates to your company officer that he is responsible for operating within the approved directions and procedures and is not to modify such guidance documents on his own.

5. What is your last-resort action?

A last-resort action might be to advise your company officer that the assigned locations for equipment on the new apparatus are established and are to be maintained as assigned, and are not to be moved to suit individual preferences. Inform your officer that if he and his crew do not comply with this requirement, you will exercise your ability to reassign the entire company to another station that has an older rig. This option is a last-resort action because it involves numerous people, some of whom are at another station with no issues—although it is likely that if the entire company is moved together, without shift changes, another company would be willing to move to a station with a new ladder truck. In addition, this option will have ripple effects throughout the department, turning an equipment placement disagreement into a high-profile, high-impact issue. An equipment placement disagreement should be resolved at the company level between company officers. It is worth reminding your company officer that you expect him to work through such issues cooperatively and to attempt to find a solution without involving any shift commanders.

6. If you reach your last-resort action, what effect will you be dealing with two weeks after taking action?

Some crew members will likely be very upset if they are involuntarily transferred. Some officers on the preceding shift, in the station, and at the shift commander level will probably be very pleased, because they will know you found a definitive solution to the problem. Some crew members may also be very pleased to be transferred from another station into Station 3 with its brand-new ladder truck. If you have a multiple-station operation, the shift commander has the ability to transfer a crew who *wants* to go rather than *has* to go. For smaller operations with only one or two stations from which to choose for the transfer, this option may not be advisable if it creates more problems than it solves. With so few stations and personnel from which to choose in making the transfer, you risk having more personnel upset about the transfers than the equipment placement. You have not really solved anything if you create more widespread discontent; instead, you have merely changed the topic of discontent within the department. For a smaller department, the last-resort action may simply be a message from the battalion officer delivered in a more forceful tone that equipment placement is not to be changed for any reason.

6. Turnout Gear Assignment

▸ NFPA 1021: 5.4.3, 5.4.5

Your deputy chief gives you the assignment to research new turnout gear options. The department historically buys this gear from what is regarded as the chief's favorite brand. After hearing the complaints of the line fire fighters following responses and training evolutions when the turnout gear is actually in use, you know that the field personnel do not really like that brand. You have to give a recommendation based on line personnel's input in one week and must determine which steps to take and how to present your research.

1. If you do nothing, what will happen?

By doing nothing, you fail to follow the order issued by your supervising officer. The task of researching turnout gear is an assignment, not a suggestion; doing nothing could result in negative consequences for you as an officer. Further, inaction could be a missed opportunity to effect positive change in the field through the personal protective equipment provided to your people and the responders throughout the organization. Even if there is a historical or perceived "favorite" brand, you are obligated to complete the assignment by objectively researching the available products.

2. What do you say or do?

First, gather the turnout gear specification standards to identify performance objectives, such as thermal performance indicators. Second, consider polling a selected sample of personnel to determine which qualities or features they like and dislike, and which functional improvements they may want to see. Be careful as you poll field personnel that you do not overpromise or commit to a particular outcome. Third, consider meeting with the department leadership responsible for the former purchase to learn about why the particular brand was purchased.

3. With whom do you address the issue?

Speak with multiple vendors to request product literature and objective performance indicators for specific brands. Inform vendors that you are not speaking exclusively with them, thereby ensuring there are no misunderstandings or perceived vendor preferences. Next, consider which field personnel are vocal in support of, or opposed to, the existing personal protective equipment. It is important to identify both points of view and collect feedback from a representative sample of personnel across multiple shifts, stations, and locations.

4. What are two other alternatives you could use?

A classic alternative is to form a committee. You have received this assignment from your superior officer; therefore, there is a legitimate need to gather input. Forming a committee of interested personnel to provide you with information can be helpful. Sometimes, however, forming a committee can be a barrier to productivity if you do not manage the project closely. This alternative also creates the risk that the committee may take discussions into the field too early or give false indications of outcomes. The committee must understand its role: You have been tasked with giving your supervising officer a recommendation, not a committee report, so ensure that the committee members recognize that their purpose is to provide you with information to make an informed recommendation.

A second alternative might be to contact comparable departments in the region and determine which information they used to make their personal protective equipment purchases. Be careful not to base your recommendation and the subsequent equipment purchase decision exclusively on low bid, a personal or familial connection to the vendor, or a desire to emulate the appearance of another fire department.

5. What is your last-resort action?

The last-resort action might involve allowing your research process and recommendation to be influenced by less than ideal criteria, such as lowest cost if funds are limited, or fastest turnaround time from order to delivery if the need for personal protective equipment is urgent. Your research might produce the most correct choice for these given circumstances or constraints, but not necessarily the absolute best recommendation overall. Remember, as an officer, you have to assess current and future situations and thoughtfully make the best choices that will yield the greatest good with the resources available within the given current time frame.

6. If you reach your last-resort action, what effect will you be dealing with two weeks after taking action?

You will likely face a firestorm of concerns that responder safety and the safety of the public whom those responders protect are compromised because of budget constraints or convenience. If you ask personnel for their opinions and then opt not to give their responses fair consideration, your credibility as an officer and possibly as an individual may be compromised in the eyes of many members of your team. Officers have the unique challenge of trying to figure out how to communicate that input received was considered and appreciated, even if it did not support the final product choice. Be prepared to explain the facts that you considered and their effect on your recommendation. You may not be able to disclose every detail considered, but you can certainly give crew members enough information to allow them to see that you genuinely considered their input.

7. Citizen Complaint: Speeding Apparatus

▸ NFPA 1021: 4.3.2, 5.2.1

You receive a phone call from a member of the public. The caller reports that earlier today he saw one of your fire apparatus speeding through an intersection just as he was about to pull out onto the street. The caller says the apparatus "could have killed somebody."

1. If you do nothing, what will happen?

Doing nothing in a scenario such as this could be considered neglect of duty. An allegation involving a piece of fire apparatus that "could have killed somebody" merits a significant investment of time and attention delving into the facts. An investigation does not imply that the allegation is credible or that anyone in the fire department is guilty, but as a battalion officer, you have a responsibility to investigate. If you do nothing, and the allegations prove to be true, the situation could repeat itself in the future with potentially more catastrophic consequences.

2. What do you say or do?

First, acknowledge that the complaint has been heard and understood. Express your appreciation to the citizen for investing the time to notify you of the situation, which you otherwise would not have known about. Give the citizen an opportunity to leave his or her contact information so that you can reach him or her if you have further questions. Also ask if the caller would like you to follow up with him or her after you have looked into the matter. This gesture reinforces the message that you are not dismissing the complaint or covering up for fellow fire fighters.

Some complainants may not want a follow-up and will be satisfied with simply reporting the incident. They may assume that nothing will be done to investigate or correct the perceived injustice they witnessed. Thus, when you offer to follow up with the caller, you implicitly communicate that you share a common value: to make things right. Remember, your concern does not mean that you have accepted that there was wrongdoing by any fire fighter or apparatus operator; it simply means you will look into the matter.

Next, gather the facts from the complainants, including the date, time, and location of the incident; the apparatus's direction of travel; and any observed apparatus markings. You will need this information to determine which apparatus operators might have been involved.

Then, research data such as dispatch logs, daily activity logs, apparatus data recorder logs (if apparatus is equipped with these logs), and other records that might reveal which fire apparatus was in the area during that time frame and under which circumstances. Finally, you will need to talk with personnel involved.

3. With whom do you address the issue?

When listening and responding to the complainant, you speak directly with him. When researching data sources, you will need to communicate with dispatchers and records administrators or similar personnel to gain the necessary information. Finally, you will need to speak with field responders. This step is where the distinction between the company officer and the battalion officer becomes apparent. As company officers, we tend to gather our crew and obtain our facts from the members, apparatus operator, or fire fighters directly. As a battalion officer, however, you must allow your company officers to carry out this task on your behalf. You may need to get your facts directly at some point, but your initial line of communication should be with the pertinent company officer(s). If you have a question about the ability of your company officer to obtain accurate facts, then you have another issue to tackle—a separate issue, to be sure, but one that you also have the responsibility to address. Do not given into the temptation to take a shortcut and gather your information without involving your company officer.

Finally, you need to keep your supervising officer informed of the situation, including the steps you are taking to address the allegation. Since the complainant is a member of the public, this individual may potentially share the same story with the media. If this were to happen, you would not want your supervisors to learn about the alleged wrongdoing from a news story. For issues with high-profile potential, you will be well served to inform your chain of command about what you have learned and what you are doing about it.

4. What are two other alternatives you could use?

One alternative might be to visit the location of the alleged incident. Examine road conditions, and identify stores or other facilities that might have been open at the time of the incident and whose security cameras might have captured the event on video. Investigate whether any of the shop attendants who were working on that day and at that time might have noticed the speed or maneuvering of the fire apparatus during the incident. If camera footage is available, request to review it. Although speed may be difficult to judge on camera, maneuvering may be visible and enable approximation of the apparatus's speed.

A second alternative could be to start the investigative process by requesting written accounts of the responses from all of the companies involved in the dispatch at that time and date. You can make this request through your company officers by providing initial information about the citizen complaint and stating that you are simply trying to gather pertinent information. By not revealing too many details at the outset, you can corroborate the facts and focus the investigation's next steps appropriately. This step also provides firsthand account documentation for your report.

5. What is your last-resort action?

The last-resort action might be to directly interview the apparatus operator in question. This will reveal the most direct information but may need corroboration from other members of the company—specifically the company officer on the response.

6. If you reach your last-resort action, what effect will you be dealing with two weeks after taking action?

As previously mentioned, if the battalion officer directly questions an apparatus operator, especially without the company officer present, that action can send a mixed signal. First, it may suggest that the operator has already been determined to be guilty, because the battalion officer rank is handling the matter directly. Second, it may send a message that the company officer is not trusted to carry out the investigation of his company. Two weeks following the citizen complaint, a wealth of rumors about the incident are likely to be circulating throughout the department, especially if you solicited corroborating information, and the fire fighters will want to know the final outcome so the issue can be officially closed. Department members are likely to assume that the report concludes the citizen was overreacting and that the actions of the company were upheld as appropriate. Regardless of the report's findings, be ready to address the results so the issue does not continue unresolved.

Interpersonal Dynamics Issues

8. Inappropriate Comments to a Peer Officer

▸ NFPA 1021: 4.2.5, 5.2.1

A company officer, Lieutenant Scott, is poking fun at a fellow officer in another station, Lieutenant Stewart, regarding Lieutenant Stewart's dating practices. You have heard some of your line personnel under Lieutenant Scott's command make references to this teasing during station rounds. Some of the comments started when the two companies were together for training. During the training session, Lieutenant Scott learned through overhearing casual conversation between Lieutenant Stewart and another fire fighter that he had an upcoming date scheduled with a new acquaintance. Lieutenant Scott heard enough to begin teasing Lieutenant Stewart. Lieutenant Scott seems to find pleasure in making others squirm. He has gained a reputation of deliberately putting other crew members on the defensive over the past nine years with the department, even when he was a line fire fighter. Just three years ago, when he was still a senior fire fighter, he consistently harassed the new recruits assigned to his station. Any time the two companies are together at training or during the wrap-up of an incident response, Lieutenant Scott takes the opportunity to get a jab in at Lieutenant Stewart. You are not aware of any disciplinary actions documented in his file over the incidents, and if any of the crew members present is bold enough to say anything to Lieutenant Scott about his constant ribbing, he simply replies that he is just joking around with Lieutenant Stewart.

1. If you do nothing, what will happen?

There are two important elements to this scenario. First, the issue is not isolated to one station, which means that knowledge of the issue has been spreading throughout the department. Second, the issue is between officers, who, because of their rank, are held accountable to a higher standard in the treatment of both their subordinate personnel and their fellow officers. Officers are responsible for preventing such situations from occurring among those personnel whom they supervise. If you allow such remarks to go unchecked, the situation is likely to escalate. You have a responsibility to act on behalf of Lieutenant Stewart, to coach Lieutenant Scott's behavior, and to maintain a positive and productive climate among personnel throughout the department. You also have a personal stake in taking action because, as the supervisor of these officers, given your knowledge of Lieutenant Scott's comments, you could become part of a harassment claim if you fail to act appropriately and address the harassment from one of your subordinate officers.

2. What do you say or do?

First, as with any issue, gather your facts. Make sure you know what was said, who said it, who was the target of the remarks, when the person said it, where he or she said it, and how the remarks were received by the target of the comments. Do your due diligence to ensure this incident is not an instance of two good friends joking with each other and having their interaction be blown out of proportion. However, as the supervising officer, even if it was a case of two friends simply joking with each other, you still need to address the fact that dating and personal choices ranging from hairstyles and hygiene issues to personal finances and sexual orientation should not be discussed at work at the fire station. Both parties in this interaction are officers, and there is a clear expectation the officers will enforce and maintain a professional atmosphere for all of their personnel. When department members see one officer joking about another officer's dating life, that behavior raises doubts regarding the ability of these same officers to enforce a professional atmosphere for their people, even if their fire fighters have not voiced their concerns.

If you find the comments were not spoken under a mutually friendly pretense, and possibly even in Lieutenant Stewart's absence, then you must go to the source, Lieutenant Scott, and address the inappropriate behavior quickly and definitively. You should act within one to two work shifts of having collected the needed background information. Definitively address the situation by holding a private conversation with Lieutenant Scott in which you communicate directly that the topic of fire personnel dating practices is off limits for discussion, jokes, implied comments, and any other communication. The focus in the fire station needs to remain on the fire department's service to the community and the preparations for such service.

Document your conversation. There must be no doubt of the time and date when you clarified your expectations. Such documentation does not have to be disciplinary in nature depending on the policies of your department and the demeanor of the discussion. The documentation can serve as a safety net for all parties involved; that is, should the topic come up again in the future, the officer will understand that your expectations were clarified previously, and the documentation will identify those expectations for which the officer will be held accountable. Such documentation also protects the battalion officer by indicating that he took action previously and discussed and recorded his expectations regarding future behavior. Claims of unclear expectations or ambiguous direction from the battalion officer will not hold up to scrutiny when detailed documentation of the conversation regarding expectations exists. If Lieutenant Scott does not seem to take the clarification session seriously or if you sense resistance, you may have to escalate the documentation to a counseling form, but this should not be your initial intent.

3. With whom do you address the issue?

Communicate directly and privately with Lieutenant Scott, the officer making the jokes. Since Lieutenant Stewart, the subject of the jokes, is aware of what was said, it is appropriate to have a private discussion with him as well. Inform him that you have advised Lieutenant Scott that such conversations are not appropriate, especially among officers, and that you have clarified and will enforce prohibitions against such conversations.

This issue is similar to arriving at a structure fire ready to backdraft. If you correctly size up the initial indicators upon arrival and take appropriate actions before charging in, then everything works out well. Conversely, if you fail to size up the situation properly and instead follow the usual routine, the result will be explosive. In this station issue, the joking is comparable to pushing smoke at the fire scene. If you treat it routinely and leave it unaddressed, it can likewise result in an explosive personnel problem, negative department publicity, harassment charges and legal proceedings, and countless hours of time spent on issues far removed from providing emergency responses.

4. What are two other alternatives you could use?

One alternative might include having an informal meeting with the crew doing the joking. If you learn Lieutenant Scott made the comments about Lieutenant Stewart's dating practices in front of his crew, then you may opt to address the expectation of what is appropriate and what is inappropriate with the entire crew present. Tell them that employees' or volunteers' personal choices, on topics ranging from hairstyles, hygiene issues, and personal finances to even sexual orientation, belong to the individual alone and that they are not to discuss others' lifestyle choices in the workplace (or in the case of volunteers, in the fire station). Encourage them to refrain from commenting about other people's choices anywhere, but acknowledge that you are responsible for and have the ability to direct and enforce fire department activity. The fire department needs personnel to focus on the performance of their duties to serve the public.

If you choose this option, an important point to remember is that you should not reprimand Lieutenant Scott in front of his crew. If you decide to talk with the entire crew, you must maintain an informative demeanor clarifying your expectations for all the members, rather than calling out the specific officer's inappropriate comments to all of his crew.

A second alternative might consist of a general update via a shift memo or newsletter article from you, the battalion officer, that reminds everyone of which discussions are generally inappropriate in the workplace—dating practices and choices, as well as personal situations including sexual, financial, or social activities, do not contribute to the response readiness of the department and, therefore, should not be topics of discussion in the fire station. This does not mean friendly co-worker conversations focusing on a brief report of a vacation destination or an upcoming ballgame is prohibited. A good rule of thumb is that if the topic is controversial or highly personal and would be inappropriate to discuss in a public crowd, then it is not a topic that should be discussed with co-workers in the fire station. These topics can lead to distraction and quickly create volatile situations in which a comment can easily go from friendly banter to offensive. Encourage fire personnel to focus on the performance of the duties to serve the public.

5. What is your last-resort action?

The last-resort action might be reserved for Lieutenant Scott, should he fail to refrain from his joking after your initial conversation. As a battalion officer, you are responsible for making sure your personnel are equipped and have clear direction on their duties so that they can accomplish the goals of the department. Creating a productive workplace is the responsibility of everyone, but supervising personnel and enforcing and protecting the integrity of the workplace are among the many responsibilities of an officer. Communicate to Lieutenant Scott that his failure to take such direction seriously will leave you little choice but to proceed with finding a replacement officer capable of fulfilling his or her responsibilities and following direction from supervising officers. Explain to Lieutenant Scott that you are responsible for ensuring that company officers can establish and maintain a safe, nonharassing, and professional work environment. If you cannot rely on him to accomplish this goal, then you must take steps to find someone who can. Proceeding with a search for a replacement is not the desired action, but Lieutenant Scott must understand the seriousness of these station responsibilities and embrace them. The choice rests with Lieutenant Scott and his words and deeds.

6. If you reach your last-resort action, what effect will you be dealing with two weeks after taking action?

Ideally, Lieutenant Scott will heed your comments and change his behavior. Even if the desired change occurs, rest assured that this incident and your response will not go unnoticed. The crews will probably think you are out to get them or Lieutenant Scott. Many fire fighters will perceive the situation—or at least the version circulating among the uninvolved personnel—as your plot to exert power by threatening to replace their good officer. They will magnify every good trait Lieutenant Scott possesses and minimize or overlook any imperfections. You will have to remain focused on your responsibility of communicating the department's need to maintain a professional workplace free from harassment, fear, or pressures regarding personal life decisions. Although many may not understand your action, your consistent message will be solid ground for you to stand upon while you steer others on the proper course.

9. Inappropriate Comments to a Fire Fighter

▸ NFPA 1021: 4.2.5, 5.2.1

A company officer, Lieutenant Scott, is poking fun at Andy, a fire fighter in another station, regarding Andy's dating practices. Specifically, Lieutenant Scott comments on the women's names, hair colors, and interests and even the restaurants Andy allegedly selects for his dates. Lieutenant Scott sarcastically instructs Andy regarding how to improve on his choices in women, the venues to frequent with his dates, and ways to "land a wife." You have heard some of your line personnel under Lieutenant Scott's command refer to his remarks during station rounds. The comments started when Lieutenant Scott's company was attending training with Andy's company, and Lieutenant Scott overheard part of a casual conversation among Andy and some of Lieutenant Scott's fire fighters—enough to begin chiming in with his self-perceived expertise on dating matters for Andy.

Lieutenant Scott has a reputation for frequently finding pleasure in making other fire personnel uncomfortable. He has demonstrated this behavior throughout the past nine years with the department, even when he was a line fire fighter. Just three years ago, when he was still a senior fire fighter, he was known to constantly pick on the new recruits assigned to his station. Lieutenant Scott pokes fun when the two companies are together at training or during the wrap-up of an incident response; any time he can get a jab in at Andy, he does. If any of the crew members present is bold enough to say anything to Lieutenant Scott about his ridicule, he replies that he is just joking around with Andy.

You are the shift commander supervising Andy's lieutenant and Lieutenant Scott. You are not aware of any disciplinary action in Lieutenant Scott's file regarding the historical incidents.

1. If you do nothing, what will happen?

The key element in this scenario is that the officer, Lieutenant Scott, is picking on a subordinate, Andy. If you do nothing, your inaction compromises the integrity of the leadership team because fire fighters rely on the officers to lead, guide, and, to some degree, protect them.

Should you choose to do nothing, you will not only implicitly condone inappropriate behavior but also risk contributing to the creation of a distrusting climate toward fire department leadership in general.

Officers, because of their rank, are held accountable to a higher standard in the treatment of both subordinate personnel and one another. They are responsible for preventing such situations from occurring with those personnel whom they supervise. If you allow such behavior to go unchecked, the situation is likely to escalate. You have a responsibility to act on behalf of Andy, to coach Lieutenant Scott, and to maintain a positive and productive climate among personnel throughout the department. You also have a personal stake in taking action because, as the supervisor of these officers, given your knowledge of Lieutenant Scott's comments, you could become part of a harassment claim if you fail to act appropriately and address an officer's inappropriate comments to a member of a subordinate rank. Andy may be reluctant to speak out against a lieutenant if he is uncomfortable or feels threatened. An officer must advocate for his or her personnel. Lieutenant Scott has this obligation to Andy, but as battalion chief, you have this obligation to both the lieutenant and Andy.

2. What do you say or do?

Deal swiftly and directly with this issue. Schedule a meeting with Lieutenant Scott and his immediate supervisor. Make sure you gather all the facts, including what was said, when it was said, to whom was it said, how often and in which context it was said, and where it was said. Make notes of the answers so specific details can be recalled later if needed. If the reporting source, such as a peer fire fighter in Andy's company, is able to indicate a possible reason why Lieutenant Scott might make such comments, note that as well. As you conduct your inquiries among anyone in the station or in any other stations who would have knowledge of the comments Lieutenant Scott made, you must indicate that you take the issue seriously, but be careful not to be overly critical or attack the parties involved in the problem. This balance can be challenging to achieve, so think carefully before you start gathering your facts.

3. With whom do you address the issue?

First, talk to the source of your information. Ask the balanced questions mentioned previously about who, what, when, where, how, and, if known, why. This conversation may produce a lead in the form of another person whom you can ask the same questions to confirm details. Then schedule a meeting with Lieutenant Scott and his supervisor. Ask the same who, what, when, where, and why questions of Lieutenant Scott. If another battalion officer is present at the meeting, explain that you have a duty to investigate such matters that involve officers and statements that could be perceived as inflammatory. Clarify that you are not accusing Lieutenant Scott of anything, but note that the supervising battalion officer needs to be aware of the information, questions, and responses firsthand so there is no misunderstanding.

4. What are two other alternatives you could use?

One alternative might be scheduling a meeting with Lieutenant Scott to conduct a direct review of the department's harassment policy and discrimination policy. You, as his supervising officer, can use this opportunity to establish the clear expectation that any comments suggestive of discrimination or harassment or discussions about other members of the department on these issues will not be tolerated and will draw immediate corrective action from the officer to protect and preserve the good order of the department. After such a review and statements of expectation, you would then ask Lieutenant Scott if he is aware of any infractions that are in need of attention. Allow Lieutenant Scott to come clean with you about his comments and past behavior. If he does not mention it, inform him directly that you have learned that he has made such comments about another fire fighter, and ask him to elaborate on the circumstances for you.

If you choose this alternative, be very careful not to make your officer feel trapped. The objective is not to ambush Lieutenant Scott with this allegation and make him feel as if he is under attack; rather, the point is to draw attention to the importance of the officer's conducting himself in conformity with department policy. An officer must not participate in conduct that is destructive to fellow members and officers' leadership integrity. Instead, officers are responsible for enforcing conduct consistent with these policies and should not in any way participate in others' infractions.

A second alternative might be assigning Lieutenant Scott a project to assess the climate on his shift or within the organization regarding awareness of and compliance with the department's policies regarding discrimination and harassment. Ask him to conduct a training session on these policies for your shift, and inform him you will be in attendance to ensure everyone understands the importance of these particular policies. You might even inform Lieutenant Scott of the allegations that he has made comments that violate these policies, but before going any further in investigating the matter, you want to ensure that everyone is aware of the discrimination and harassment policies and their application. This alternative allows you to address the issue indirectly with Lieutenant Scott and offers the added benefit of reminding everyone in the department of these requirements and the importance of following their guidance. If you pursue this option, you will still need to review the allegations and your clear expectations for Lieutenant Scott regarding appropriate behavior. Do not allow this issue to be brushed aside with indirect action only.

5. What is your last-resort action?

This scenario is complex and can be addressed successfully in many ways, especially considering that departmental policies can be very diverse in both guidance and consequences when it comes to the subjects of discrimination, harassment, or officer conduct. You may have already thought of some ideas that are not covered here. Typically, departments do have some policy in place prohibiting harassment and discrimination or mandating officer conduct standards that prohibit speaking badly of another member of the organization. The last-resort action would likely be for you, as the supervising officer, to bring Lieutenant Scott up on disciplinary charges that are formally filed through the chain of command. This is the last-resort action because it skips the education and corrective steps for your officer and moves directly to the punitive stage, which can quickly become severe. This alternative may be necessary depending on what was said, how and where it was said, how it was intended, how often it was said, and other factors. At the same time, it opens up the discovery of and debate over which details are accurate, the sources from which the details were learned, and so on. It puts the inappropriate comments on center stage for all to see, which can be detrimental to Andy, the subject of the comments; Lieutenant Scott, the officer; and the department as a whole. Charges may be merited and can be considered. Recognize, however, that once charges are filed, the consequences are nearly impossible to undo, even if the charges were to be later dropped.

6. If you reach your last-resort action, what effect will you be dealing with two weeks after taking action?

You and everyone else in the department will be dealing with a formal investigation. Rumors and hearsay will run rampant, and members will begin to align themselves on the various sides of the issue. Filing a formal complaint is not likely to be a quiet or simple process. Andy, who was the target of the harassment, may just wish the matter would go away. Of course, once charges have been filed, this outcome is not possible. In these situations, it is important to discern the true objective of your action—namely, all members, especially officers, conducting themselves in compliance with the departmental policies and engaging in harmonious work relations. Many paths can be followed to reach this end, each of which has its own advantages and disadvantages. Size up the situation and take careful action.

10. Fire Fighter with Skill Deficiency

▸ NFPA 1021: 4.2.3, 4.2.6, 5.2.1, 5.2.2

You are a shift battalion chief in a combination department, and keeping career and volunteer members on the same page is a constant challenge. A career member, Jill, reports to you that one of the volunteer members, Dan, has not mastered basic fire fighter skills, such as breathing apparatus emergency procedures, hose stream selection and safe use, and even fire behavior recognition—all of which are necessary to perform successfully as a fire fighter. Jill informs you she does not trust Dan to work alongside her on emergency calls. She tells you that she has reported her concerns to her company officer, but nothing has been done about it, and Dan is still responding to calls.

1. If you do nothing, what will happen?

If you do nothing, things will get worse. An allegation of skill deficiency is a serious issue in need of investigation. There is also an allegation that your officer has not addressed a serious allegation that was brought to her attention. If you do not act, that choice will compromise your integrity and your ability as a leader to supervise your officers; it will communicate indifference and the notion that it is acceptable for unskilled personnel to participate in emergency responses and possibly compromise the safety of all responders involved. Taking action does not indicate that either allegation is true; however, to confirm their validity, you must engage in the process.

2. What do you say or do?

First, respect and work through the chain of command. You have received word from the fire fighter, Jill, that her company officer has been informed of the situation; start there. Find out what the officer knows and which (if any) actions she has taken up to this point. Just as it takes discipline to walk around a burning building to size it up rather than grab a hose line and get water on the fire, as a battalion officer, you must call upon the same discipline to work through your company officer. Size it up; if your officer has initiated actions, support her. Do not be tempted to jump in and resolve the issue yourself. However, if your officer is not taking appropriate action, or even no action, do not bypass her but rather use this opportunity to coach her in which steps to take and why her action is necessary.

3. With whom do you address the issue?

Communicate with the company officer. Unless she reveals to you in that discussion some legitimate reason why she is not the most appropriate person to handle the issue, you should encourage her to address this situation at the company level. It may be appropriate to communicate to Jill, the fire fighter who raised this issue with you, that you are communicating with her company officer about the matter. Let her know you did not simply ignore the information, but be cautious not to imply you are handling the matter directly. You do not want to erode the integrity of your company officer with her crew.

4. What are two other alternatives you could use?

One alternative is to conduct a skill drill for all of your companies. This drill will allow you to witness firsthand whether Dan, the volunteer fire fighter in question, demonstrates deficiencies with basic skills. It will also enable you to assess these skills for all of your personnel—not just the one fire fighter brought to your attention—and can be beneficial to you as a shift commander. If your observations indicate that Dan may truly be deficient in basic skills, you should still assign the company officer to handle any corrective actions—either skills or discipline. The company level is the most effective level at which to assess, practice, and encourage skill development and performance. As a shift commander, you must work through your people, the company officers, to accomplish department skills.

A second alternative might be to design a skills competition to be conducted among your shift members or across the entire department, depending on your organization's size. Design a competitive forum for individual performance and recognition. This alternative would typically require a longer period of time to carry out, approximately three to six months. Again, be careful that you do not go to elaborate extremes to address a possible situation with one member. If Dan does, in fact, have a skill deficiency, then coach Dan's company officer to address Dan's deficiency within a defined and short period of time, rather than devoting large amounts of time and energy to making everyone on the shift participate in an exercise that is intended to address an isolated deficiency. Such a skills competition can actually be beneficial to the entire organization if creatively designed. If the competition is a snare, however, it may be recognized as a snare, even by the high performers within the department. In that case, you run the risk of creating new problems that must be addressed, such as distrust and even performance lag.

5. What is your last-resort action?

The last-resort action would be to discipline your officer. If Dan does have skill deficiencies and the company officer was informed of the deficiencies but failed to take any action in a timely manner, then you will need to counsel the company officer regarding her performance.

6. If you reach your last-resort action, what effect will you be dealing with two weeks after taking action?

There are many possible outcomes from the step of counseling the company officer. Some career members may be pleased that you have engaged with the officer to provide a safer work environment for them. Some volunteer members will likely feel unjustly singled out and held to an unreasonable performance standard—not just Dan but many or all volunteers. The officer will likely be upset that Dan's lack of performance has created a blemish on her leadership.

As a shift commander, you may be tempted to step back from the issues that develop in this scenario and let them be addressed at the company level. No matter what happens, however, you have a responsibility to coach and lead your people. As a shift commander, your responsibility is to develop your company officers to recognize and embrace the importance of honing their skills—to make them better leaders of their companies, not just in emergencies but in all situations. It is your responsibility to clearly define the expectations you have for your company officers in regard to addressing station problems, attitudes, disagreements, and skills deficiencies. The company officers can keep a keen eye on which issues are brewing and correct the course before problems evolve into full-blown crises. Do not solve the issues for them; help them recognize the importance of becoming effective problem solvers for the good of the department.

11. Career Member Questions Volunteer Members' Skills

▸ NFPA 1021: 4.2.5, 4.2.6, 5.2.1, 5.2.2

A long-time volunteer member, Frank, files a complaint with you, the battalion chief, regarding a recently hired career member, Tony. Frank believes that Tony is arrogant and condescending, and he is concerned that Tony's attitude is negatively affecting the company. Frank's written complaint describes that Tony "struts through the bay" with the other volunteers in attendance for training and that he is often heard telling the volunteers that if they work hard, they, too, can someday be hired and "learn how to be a 'real' fire fighter—not just play fire fighter when a call comes in." Frank reports that he has discussed the issue with the station lieutenant, but nothing appears to have been done.

1. If you do nothing, what will happen?

If you do not act, you allow the tensions to escalate, which is a recipe for disaster for the individuals involved and the organization as a whole. Your failure to act also sets your company officer up for failure. If you are aware of the situation and you do not communicate it to your company officer, you leave the company officer with a challenging issue to deal with on his own when the situation erodes. Note also in this scenario that the volunteer fire fighter, Frank, "filed a complaint." If the complaint is filed, it means you are now officially aware of it and responsible for its disposition. You do not necessarily have to personally deal with the situation, but it cannot be dismissed without implicating you as well.

2. What do you say or do?

First, you must investigate to determine whether Frank reported the issue to the station lieutenant before he brought it to you. The organization has a chain of command, which should be both respected and used. You must direct the complaint to the appropriate person to preserve the effectiveness of the communication process within the organization's chain of command.

Upon investigation, if you determine that none of the company officers is aware of the complaint, that lack of awareness presents an educational opportunity for Frank, your volunteer member, to learn about proper procedures. This education is probably handled best by one of the officers identified in the organization as closest to Frank in the organizational chart.

If you determine that one or more officers are already aware of the issue from Frank, seek information about what they have done up to this point and what they plan to do in the future. Explain to the officer(s) that you were made aware of the issue by a jump in the chain of command but that the breach of the chain of command's communication channels does not indicate that you will become involved any further than participating in a briefing to ensure your situational awareness. Maximize the opportunity to build confidence in your company officers while still communicating the importance of the chain of command to the member.

In this scenario, you, as the shift or battalion commander, have yet to take ownership of the situation. You do not know whether the complaint has merit or whether any follow-up action is appropriate. This point is important because not all issues that make it to the battalion commander are appropriate for the battalion commander to own. Action is appropriate, but it might simply consist of gathering intelligence on the matter and making sure the company officers are aware and are acting on the issue. If you become involved with every issue that jumps the chain of command, you will quickly find yourself out of time and energy to handle the responsibilities that are part of your position. This type of distraction can be compounded when dealing with a volunteer member, because career versus volunteer attitudes and cultures can erode quickly if either group feels their perspective or issues are being overlooked.

3. With whom do you address the issue?

First, ask Frank, the volunteer member, to identify the company officer with whom he spoke before coming to you. If his answer is vague or nonspecific, remind Frank of the importance of the chain of command, and refer him to the appropriate company officer to handle the issue. Second, speak with the company officer(s) identified. If none was identified, determine who should be made aware and contact him or her to provide advance notice of what should be coming the officer's way and from whom.

4. What are two other alternatives you could use?

One alternative might consist of a reprimand of Frank for operating outside of the chain of command. If Frank had previously communicated with a company officer and came to you without the company officer's knowledge, then Frank was either side-stepping or attempting to overturn a company officer's previous action. There is an outside chance that Frank brought his complaint to the battalion officer because his previous communication with the company officer resulted in no action at all; therefore, he is not circumventing or trying to overturn the company officer's decision. Even if this were the case, Frank must still understand that he has an obligation to inform his company officer that he wishes to have an audience with the battalion officer to discuss the complaint against Tony. The battalion officer has another issue to address if the company officer has received a complaint and has not acted upon it—but that is a different scenario from the one considered here.

If Frank did not inform the appropriate officers of the situation, then the infraction would be that he failed to bring the issue to the appropriate company officer before taking a complaint to the battalion commander. A rebuke for this failure will certainly not sit well with the volunteer member, and possibly with other members as well. Nevertheless, the chain of command must be respected, especially at the battalion commander level of communication.

A second alternative might be to make increased visits to the company in question. By increasing the frequency of your visits and focusing your attention on both the career member, Tony, and the volunteer member, Frank, you will gain firsthand knowledge of what is really occurring. Again, this step raises your situational awareness. Note that observing the interactions and skills does not create battalion commander–level ownership but rather allows you to assess whether either Frank or Tony is telling tall tales.

Even if you implement the focused situational awareness alternative, it is important to communicate your observations to the appropriate company officer whom you identify as responsible for addressing the situation. Be sure to have the company officer address the situation and document the findings and outcome. Although disciplinary action may not be appropriate, communicate to the company officer that you would like to receive a fact-finding memo addressed to you that documents the topic, the individuals involved, and the findings. Such documentation is important should future events ever arise. It is also important for the company officer to understand the importance of report writing and documentation of any personnel matters with which he or she becomes involved, regardless of how seemingly insignificant they may appear.

5. What is your last-resort action?

Your last-resort action is to become actively involved in the investigation. Determining the "who, what, when, where, and why" of the events will consume significant time. Having a battalion commander involved in the situation adds to the impact because of the positional responsibility associated with this level of authority. Finally, battalion commanders have other roles to fulfill that have organizational effects far beyond clashes of two individual fire fighters' attitudes. Company officers are positioned to observe, interact, and correct fire fighters (career or volunteer) more responsively and rapidly than battalion officers. To have the battalion officer investigating attitude issues instead of the company officer would be a last-resort action by the battalion officer.

6. If you reach your last-resort action, what effect will you be dealing with two weeks after taking action?

You will likely fall behind in managing your other battalion duties. Company officers may become resentful if they perceive that you conducted the investigation yourself because you do not believe they have the capability or trust them to complete the assignment themselves. You may find that the outcome will be increased dissension within the organization. Regardless of what you determine to be the appropriate response, one side may feel wrongly attacked and defensive, and the other side will likely feel ignored or devalued. Tensions will be high, and all officers will need to maintain heightened awareness to prevent overreactions and flared tempers.

12. Negative Feedback About a Superior Officer

▸ NFPA 1021: 5.4.5, 5.4.6, 6.4.6

Chief Willis, the operations chief, assigns you the task of assessing morale within the department and reporting your findings to him, along with suggestions for improvement. He gives you no further direction, as is often the case at a battalion command level. Through your face to face discussions during regular station visits and brief survey of other shifts' personnel who are not readily accessible for face-to-face meetings with you, you receive overwhelming responses from personnel suggesting that Chief Willis is the source of many of the department's problems. The shift personnel perceive that Chief Willis is out of touch with the line and inconsistent in his decisions affecting the department membership. Lack of respect for and inconsistent or even unfair treatment of line personnel is becoming more commonplace. Personnel cite many specific examples to justify their feelings, so you are fairly certain the unhappiness is not just a matter of the personnel not "liking" Chief Willis. Furthermore, equipment and facilities are aging, and the line fire fighters believe that Chief Willis is not lobbying vigorously for needed improvements; they interpret this situation as Chief Willis being unappreciative of the hard work they accomplish at the street level of the department.

1. If you do nothing, what will happen?

This one is tricky. If you do nothing, you run the risk that Chief Willis will counsel you for not completing an assignment. If you do not report the findings, and he does not ask about it, you might think you can escape the unpleasant task of having to report to Chief Willis that he appears to be the perceived problem. Even if Chief Willis fails to follow up on his assignment, however, the people whom you have polled will be keenly aware that you sought their input and they responded by identifying a problem—and now you must act. Your people are depending on you to complete your task. Doing nothing would jeopardize your relations with your shift and could very well exacerbate the department's problems.

2. What do you say or do?

First, gather the responses you received and group them if possible. Categories of responses might include operational/tactical concerns, communication problems, trust or fairness issues, and maintenance and repair issues. Specific categories of responses would be based on the actual responses received.

Next, document your findings in a written report that summarizes the findings by category and by issue; if particular responses were frequently repeated, quote the particular wording used. It is critical that you keep all comments and responses anonymous. Some might argue that the comments are valid only if they are attributed to specific respondents. This is where you must consider your assignment. You were not asked to identify specific people and their issues; you were tasked with assessing the department's overall morale and reporting your findings and suggestions.

Since your results revealed that Chief Willis is a significant source of discontent, provide suggestions appropriate to the feedback received with an emphasis on points of consideration for the operations chief. It is not entirely appropriate for you to recommend specific behavioral changes to the operations chief; he must come to this conclusion on his own or with assistance from his superior officers. Instead, suggestions should be limited to opportunities for the operations chief to positively impact the climate of the department in ways of which he might otherwise be unaware. You should not simply suggest to your supervising officer that he move on, retire, or step down. These actions do not fix anything and can be detrimental to your future as well.

3. With whom do you address the issue?

This information is sensitive, and the results are especially negative for the operations chief. Be true to yourself, your agency, and your subordinates by fulfilling the assignment, even if the results are hard to report. Because of the inflammatory nature of the results, record the findings in a written report that you deliver personally and make a verbal report about the document to the operations chief who requested it. Do not share results with other officers, with the shift personnel, or even with other chief officers. You do not want to be responsible for spreading negative comments about Chief Willis to others before he has been informed of the results. It would be wise to communicate such steps to Chief Willis during the meeting in which you report the results of his assignment.

4. What are two other alternatives you could use?

First, you could gather the information and submit the written report to Chief Willis without meeting with him to discuss the findings. This alternative may be easier to implement but will leave you wondering how it was received and unable to redirect the findings in a positive direction should Chief Willis respond with hostility to the unpleasant news.

A second alternative might be holding a midpoint meeting with Chief Willis to discuss some of the preliminary findings. Share that some of the early responses indicate a chief officer is contributing to the climate and morale, and then ask him if he has any insight into the situation. If Chief Willis states he is unaware of any information that might shed light on the issue, then his response indicates he has no knowledge of his role in the situation. This is important to recognize, because it will be helpful to know that Chief Willis will likely be surprised about the report results and could become defensive upon learning of his role in the department's morale situation. Should he press for information about the responsible officer, you would be in a good position to advise him that it is still too early to know for certain and that more information will be included in the final report. If he offers that he is likely the officer in question, then you can complete the assignment with insight into how he might receive the final results. A midpoint meeting also offers you an opportunity to ease into the findings rather than blindside the operations chief later with a complete report that vilifies him as the culprit. No one likes to be perceived as a detractor or a poor performer. Thus, this alternative allows for making gradual progress, and it enables Chief Willis to adjust his behavior before the final report is written, thereby potentially softening its impact.

5. What is your last-resort action?

The most inflammatory action—a characteristic that makes it a last-resort action—would be to complete the assignment by identifying Chief Willis as the root cause of the negative morale as indicated through shift feedback and by recommending that Chief Willis make significant changes to remedy the situation. Although this may sound similar to the previous options, this action adds the bold step of a subordinate officer making specific recommendations to his or her supervisor to improve the situation.

This bold step also allows you to assess whether Chief Willis is following thorough with the report findings and suggestions, which leads to the second part of the last-resort action: If Chief Willis does not take action after a reasonable period of time (as measured in months, rather than hours or days), consider requesting an audience with the fire chief to share the assignment, the process, the findings, and the lack of noticeable changes from Chief Willis. During this meeting, you can brief the fire chief, while being upfront about your inability to see whether steps are being taken at levels outside your vantage point. You do not want the fire chief to think you are accusing Chief Willis of inaction, when he may actually be taking steps of which you are not aware.

6. If you reach your last-resort action, what effect will you be dealing with two weeks after taking action?

Your supervising officer, Chief Willis, may well be displeased with you for taking the matter beyond him in the chain of command. No one wants to be reported to his or her boss. Increased tensions may also be directed at you from the shift personnel. Those personnel shared their feedback with you for the initial report but may have seen no evidence of change. Some may perceive you as part of the problem rather than part of the solution. Just as you are unsure which steps the operations chief has taken or is taking, so the shift personnel are unsure if you have used their input or edited it because they have not seen any tangible results or changes.

13. Company Officer Disregards Chain of Command

▸ NFPA 1021: 4.4.4, 5.2.1

Captain Biltz is an enthusiastic company officer at fire department headquarters. Whenever he has operational or administrative questions, he routinely bypasses you as the battalion chief and solicits the deputy chief officers and sometimes the fire chief for answers instead. These questions range from the rationale behind the company training rotation to which content might be covered in an upcoming promotion process so that he can be well prepared. Captain Biltz never brings these types of questions to you, his immediate superior. Sometimes Captain Biltz even reports the questions' answers back to you, as if to keep you informed. You are frustrated with his actions. Captain Biltz is a relatively new captain, holding the position for only a year now. His enthusiasm for the position and eagerness to obtain answers to his questions are how the questioning of the deputy chiefs originally started, but over the past three months, the frequency with which Captain Biltz's questions bypass you as his supervising battalion chief has increased significantly, occurring once or twice a week now.

1. If you do nothing, what will happen?

If you ignore the issue, the enthusiastic Captain Biltz will not learn the importance of the chain of command, and the chief officers will eventually become annoyed at Captain Biltz—and possibly you, as Captain Biltz's supervising officer. Captain Biltz will likely wear out his welcome with the chiefs, and he may even become frustrated that the department leadership does not seem to act on or appreciate his eagerness to learn and improve. Over time, the company officer will lose his enthusiasm and remain unaware of the importance of respecting the chain of command, which has the potential to become a catalyst for future negativity.

2. What do you say or do?

First, recognize the importance of the chain of command and the reality that sometimes the obvious is overlooked. Specifically, your officer may not recognize that taking these questions to the chief officers is a breach of the chain of command. Captain Biltz may have a situation or what he believes to be a simple question and communicate it to the first officer he sees or to an officer with a personality similar to his own with whom he is comfortable talking. Before reprimanding Captain Biltz for jumping the chain of command, take advantage of this scenario by using it as an opportunity to educate him about or reinforce the necessity of a chain of command.

3. With whom do you address the issue?

Talk to Captain Biltz, the company officer, directly and privately. Inform him that many questions or concerns at the company level can and should be addressed by the battalion chief. If assistance is needed from a higher rank within the organization, the battalion chief is the most appropriate officer to seek such assistance, coordinate assistance, or follow up on requests.

4. What are two other alternatives you could use?

The first alternative would be to cut the information source off from above. As the battalion chief, make a proactive request of the chief officers that, should Captain Biltz bring a question, concern, or idea to them, they simply refer Captain Biltz back to you, the battalion chief. Ask them not to provide any response to his questions, even if it is an easy answer, so you can educate your company officer on the importance of respecting the chain of command. Explain to your other chief officers that they are welcome to follow up with you whenever they are aware of an issue that has been brought to them, but explain that you would like the opportunity to decide how the request, idea, or question is handled.

A second alternative might be to educate Captain Biltz about the importance of the chain of command and its objective to solve problems at the appropriate level. Such an education session can convey to Captain Biltz how use of the chain of command actually liberates chief officers to plan, equip, and staff the organization in the most effective way, while still providing assistance to the company officer. The session can be conducted in a manner that is constructive and goes beyond the "just ask me instead of them" perspective. Such a discussion might potentially be misunderstood as an attempt to micromanage Captain Biltz, but the emphasis of the conversation should be the importance of the chief officers' responsibilities on behalf of the entire department. Captain Biltz's questions and frequent drop-in visits pull the chiefs off task to answer questions that can and should be handled at the battalion level, and this effect is counterproductive.

5. What is your last-resort action?

If educating Captain Biltz and limiting or cutting off the supply of information or assistance from the chief officers do not resolve the issue, you may need to have a direct conversation with Captain Biltz about what he can and cannot do. Inform him in a private meeting that all requests for assistance or questions outside the company must be directed to the battalion chief. Be clear that this is not a suggestion but rather a directive from a supervising officer.

You can opt to explain your reasons for issuing the directive if you wish, but the direct order must be unmistakable and unambiguous. For example, you might explain to Captain Biltz that this directive is for the benefit of everyone within the organization. Inform him that, as a battalion chief, you have a responsibility to supervise your company officers, which includes handling battalion-level questions, concerns, or requests on behalf of the department. When he bypasses the battalion chief, this responsibility is compromised. Your directive, then, is intended to help reestablish good order for the company and the shift. You can help Captain Biltz understand this rationale by changing the perspective when discussing the matter with him. Ask him how he would accomplish his responsibilities as captain if his personnel constantly went around him with their questions and concerns and how those actions might make him appear ineffective or inadequate in the eyes of his other subordinates or his supervisors, even if he was never aware of or given a chance to address the company's concerns. Then explain how this directive allows you as the battalion chief to do your job and assist your company officers when they have questions, concerns, or needs beyond their company.

6. If you reach your last-resort action, what effect will you be dealing with two weeks after taking action?

You can expect an increased number of questions from Captain Biltz. There may be some resentment at first, but if you follow up regularly with Captain Biltz and express sincerity in wanting to help, the effect can be very positive for everyone involved. By taking a direct approach, and because Captain Biltz was enthusiastic from the start, indicating a desire to perform well, you will likely experience appreciation from your chief officers for the reduced number of questions and distractions as well as Captain Biltz's gratitude that you are there to help him. This approach could even have some educational value for the entire company if Captain Biltz were to share the experience with other company officers. The most important point to take away from this scenario is that to reach a positive outcome, you must follow through with answers, assistance, and support by using a positive tone and emphasizing the good of the organization. If you fail to do so, the results may be misperceived as a ploy to subdue or control your personnel and keep them silent. This outcome can be devastating to a company, shift, and culture.

14. Fire Fighter Fraternizes with City Manager

▸ NFPA 1021: 4.2.5, 4.4.4

Your department has a chain of command that all members understand they are expected to follow. You learn from your ladder company officer at Station 2, Lieutenant Ewing, that a member of his company, Kevin, has become increasingly friendly with the city manager. Kevin did not previously know the city manager, but since her appointment and Kevin's election as fire department labor representative, the city manager has been inviting Kevin to dinner and sporting events with her. Kevin shared this information with Lieutenant Ewing one day during casual conversation during a shift weight-lifting session. During the same conversation, Lieutenant Ewing mentioned that it was nice to have fire department personnel getting some face time with the city manager outside of city hall. Kevin then revealed that in some instances, the only city personnel in attendance at these events were the city manager and Kevin. Kevin went on to say that the city manager's family and Kevin's family had also attended multiple family picnics together. The conversation between Kevin and Lieutenant Ewing during this weight-lifting session was the first Lieutenant Ewing had heard of this interaction between Kevin and the city manager. Lieutenant Ewing recognized this relationship to be highly unusual but was not sure what to do or say about it, so he pulled you aside one day when you made station rounds to make you aware of the situation. The lieutenant further informs you that based on general comments made during the lieutenant's interaction with Kevin, he believes Kevin's off-duty conversations with the city manager sometimes involve fire department business.

1. If you do nothing, what will happen?

Inaction may avoid stirring up an issue where there are limited or no legitimate grounds to discuss the events. However, doing nothing at all means failing to monitor the situation, thereby leaving you, Kevin, and the department at risk of this situation developing into a problem that falls within your scope of authority but about which you took no action. If you do nothing, the relationship and activities are likely to continue. Increased activity or even a newfound friendship with a city official does not constitute any wrongdoing, but it may test work relationships—so be on guard. Such outside relationships create opportunities in which personal and professional relationships become less clearly defined and certainly less carefully monitored by the institutions of city hall and the fire department. Even a friendly discussion about how things are going at work could quickly develop into detailed conversations regarding either party's perspectives on the workplace or workplace activities, policy decisions, or future endeavors.

What makes these types of discussions, details, or perspectives different from other relationships is the fact that the city manager has legitimate decision-making authority in matters pertaining to the fire department. Further, to maintain an effective chain of command in any organization, all communications—excluding those dealing with any immediate life-safety threats—should be conducted within the chain of command. When ideas, disagreements, or alternative points of view are shared with the city manager or a fire department insider without the fire department's chain of command being aware of such communications, that circumvention of the normal communications process can erode trust and effectiveness for both aspects of city operations: the fire department and city hall. If Kevin mentions in a casual friendly conversation at a social function that "the guys at work" do not agree with a particular policy or directive, the city manager could interpret this comment to mean that Kevin is speaking for the entire membership, especially considering Kevin's elected position as labor representative, resulting in the city manager's relying on inaccurate information when adjusting public policy.

2. What do you say or do?

The first thing you should do is pay attention. Be alert to conversations and occurrences. This does not mean you should eavesdrop or be secretive but rather that you should pay careful attention when you are carrying on conversations around the fire station. You may stumble upon an opportunity to casually and privately bring up any concerns during the course of your regular interactions with Kevin and Lieutenant Ewing rather than designate a formal time and place to address Kevin's relationship with the city manager. You must recognize that you do not get to select your subordinates—or, in this case, your subordinate's friends and associates away from the workplace.

Next, review your department's or municipality's policies to determine if they prohibit intergovernmental associations. This particular relationship is not likely to be discussed in the existing documents, but there may be relevant policies in place—for example, a policy stating that fire department members are not allowed to communicate with city officials unless they properly utilize the chain of command. In this case, one could argue that the socializing is not a work matter, so the chain of command is moot; however, as the shift commander and an officer of the fire department, you must at least investigate whether any such policies exist and learn which parameters are acceptable. If no such policies address this issue (and that may very well be the case), then you will have to press on into uncharted territory. You will have to use your best judgment to guide discussions and set parameters because there is no policy guidance available.

3. With whom do you address the issue?

You would want to investigate the existing policies first, and only then consider talking directly with Kevin. To preserve the chain of command, you should inform Lieutenant Ewing, Kevin's company officer, that you need to speak to Kevin directly regarding a non-workplace issue. You should also request that Lieutenant Ewing be present during this conversation but allow you to address the issue directly with Kevin, since the issue appears to be one that is outside the realm of Lieutenant Ewing's workplace responsibilities.

When meeting with Kevin, explain that the discussion is not about any work performance issue but rather is intended only to raise his awareness about a delicate situation. Emphasize early in the conversation that to your knowledge, nothing Kevin is doing is wrong or inappropriate but that you want to discuss the issue before a blind spot develops or Kevin finds himself in a difficult position. Explain how increased socializing can eventually lead to work-related discussions due to the common employer shared by Kevin and the city manager. It would not be uncommon for members of the same social circle to ask about each other's situations and experiences at work in any occupation. In this case, however, because the occupation (or even volunteer status for a volunteer department) is directly linked to decisions, actions, or inactions that the city manager takes during the course of her normal duties, Kevin will be in breach of the chain of command should any workplace conversations arise. Because of the nature of employment in the case—both fire fighter and city manager within the same city—and the fire department's reporting relationship to city hall—specifically, the city manager—Kevin's conversations at these social outings could easily be misinterpreted by anyone who discovered the jointly attended event as work-related discussions or attempts to influence city hall or even the fire fighters.

Although specific details of conversations may be considered to be private and Kevin may not disclose them to other parties, there is a possibility that the conversations might become known at city hall through casual comments made by the city manager, just as Kevin mentioned how he had attended sporting events and family picnics with the city manager to Lieutenant Ewing while weight lifting at the fire station. Because of the unique workplace positions of Kevin and the city manager, their interactions open the door for those who want to believe the pair's co-attendance at social events and subsequent conversations are more than just friends socializing. Just because people talk and spread rumors, it does not automatically mean there is any truth to such comments. Nevertheless, you can explain to Kevin that by placing himself in these situations, he allows the stories and insinuations to blossom.

4. What are two other alternatives you could use?

One alternative might be to inform your supervisor about this situation. Seek his or her counsel on the matter before talking to Kevin. Brief your supervisor about what you found regarding any policy direction that the department or the city/district has in place. As a battalion chief, you do not have the same access to the city manager as your supervising chiefs might have, and they may opt to have a discussion with the city manager rather than the fire fighter. It is also important that you brief the supervising officer to keep the upper levels of the fire department informed.

A second alternative might include having your company officer, Lieutenant Ewing, who supervises Kevin, investigate the situation further through routine dialogue. This option can be effective if you have a company officer who fully understands the sensitive nature of the matter and the limitations and boundaries of what can be discussed. If you are concerned about your company officer's ability to handle the situation tactfully, then you will probably want to keep the conversation between yourself and Kevin. Brief the company officer, Lieutenant Ewing, that even if he is present during your meeting with Kevin, you expect him to simply listen while you conduct the conversation with Kevin. In this kind of situation, you are operating with shades of gray rather than black or white. Shades of gray can be argued either way, and the possibility of multiple interpretations will usually raise concerns among legal counsel regarding departmental risk. To avoid arguments, especially if legal questions arise about the departmental boundaries regarding personal liberties, you should have your most tactful, human-resources-savvy fire officer to handle this delicate conversation.

5. What is your last-resort action?

The last-resort action might include seeking a legal interpretation of the risks from the department's attorney. This option is considered a last resort because if the attorney opts to take any action at all—even just a phone call or brief fact-finding visit—then the matter automatically escalates because legal counsel has become involved. There is no way to sugar-coat or down-play the significance of an attorney looking into the social relationships of a member of the fire department, because the involvement of an attorney puts the situation in a more serious light.

6. If you reach your last-resort action, what effect will you be dealing with two weeks after taking action?

The after-effects will depend largely on which action(s) the attorney takes. Other effects might include extreme displeasure from any fire fighters, including Kevin, who feel the fire department has poked its nose too far into the personal matters of the membership.

This displeasure will likely be directed toward the fire administration, including you as the battalion chief, for concerning yourself with the private lives of your members to this degree. The situation has the potential to deteriorate before it gets better, specifically with increased activity occurring between Kevin and the city manager or other officials. For example, if Kevin is approached by the attorney to discuss how such social relationships could be perceived as a conflict of interest or a compromise to the order of the city's and fire department's chains of command, and Kevin then becomes offended that the department has expressed any opinion regarding his off-duty social activities, he may very well increase the frequency or public profile of such activities just to prove a point. Should Kevin or the city manager take such a position, the outcome would be the very opposite of what had been intended.

The downside to this scenario is you may have limited actions at your disposal. There is also increased risk of offending the involved parties and further detracting from the department's mission readiness. Weigh the circumstances and the personalities of the players involved heavily before proceeding.

Sensitive Policy Issues

15. Last-Minute Arrival for Shift Changes

▸ NFPA 1021: 4.2.2, 4.2.5, 4.2.6, 5.2.1, 5.2.2

As the battalion commander, you try to set a good example for your company officers by being in your work station at least 15 minutes before shift change to make sure you can be ready to go to work the moment the clock strikes shift change time. Captain Jimenez usually does not arrive at the station until immediately before the shift change. He is often seen hastily slipping on and buttoning up his uniform shirt as he walks across the parking lot to enter the station before the clock strikes the 8:00 A.M. shift change. Captain Kirk, the offgoing captain, understands that shift change does not happen until 8:00 A.M., but he also knows that when the company is called for an emergency minutes before shift change, he must respond because Captain Jimenez is unlikely to be on the fire department property at that point. Captain Kirk, like most company personnel, understands that things happen and help is needed every once in a while, whether because of a big fire response late in the shift or a flat tire on the way to work. Captain Kirk has no issue with helping another company officer on occasion, but Captain Jimenez's just-in-time arrival has become a habit. It has created scheduling hardships for Captain Kirk and resulted in overtime for the department. Furthermore, Captain Jimenez has assigned his crew to make sure everything is ready to go before he arrives at the station, including placing his personal protective equipment on the apparatus in the officer's position and having an attendance report for his company ready as he enters the door. Both Captain Kirk and many of Captain Jimenez's fire fighters are discouraged by what they feel is Captain Jimenez's lack of respect for others. One day after shift change, one of Captain Kirk's fire fighters calls you to tell you about this situation and the frequency with which it is occurring. The fire fighter informs you that Captain Kirk deserves timely relief and that his crew is concerned that, although he is frustrated, he will not complain or risk creating tension with his peer. The fire fighters working for Captain Jimenez see what is happening but also are hesitant to raise the issue out of concern that it might appear insubordinate to their captain if they jump the chain of command, especially regarding an issue in which they are not directly involved. No one knows exactly what to do about the situation, so this fire fighter decided to call you to report what is occurring. They look to you as Captain Jimenez's officer to address the situation.

1. If you do nothing, what will happen?

The long-term impact of doing nothing will reinforce for future officers that this level of minimum effort is acceptable. The fire fighters to whom Captain Jimenez has given these directives will certainly remember this lesson, and if the captain's behavior goes unchecked, they may adopt the same behavior when they are presented with the opportunity to promote and lead a fire company of their own. If you do not address the issue, you communicate to the existing workforce that reporting for duty at the last minute and having others carry your responsibilities is acceptable—which it is not. You must engage to correct Captain Jimenez's inappropriate behavior before it frustrates Captain Jimenez's fire fighters further and compromises your leadership ability to identify and address substandard behavior.

2. What do you say or do?

The first thing you should do is objectively investigate the situation. Gather information to confirm whether you have a complete picture of the situation. Assess the dynamic between Captain Jimenez and his company's fire fighters. Determine the rationale behind Captain Jimenez's preparatory requirements. He may be addressing a substandard performance issue within his company and using this practice as a method of implementation, or he may be intentionally abusing his authority by assigning others to perform his responsibilities. Do not assume that just because Captain Jimenez is coming in at the last minute, he has made this assignment simply for his benefit.

3. With whom do you address the issue?

This is where you need to be willing to confront Captain Jimenez with a direct conversation to get his side of the story. Captain Jimenez is your subordinate, and you are responsible for supervising him. He, in turn, is responsible for supervising the company. To assess what is happening, engage in a private conversation with Captain Jimenez, and ask him about the company's procedures when reporting for duty and any problems during this time. Allow him to answer fully. After he has given you this information, it would be appropriate to share with him that you have received a complaint regarding his lack of promptness in reporting for duty, and you need to get the facts of the matter. The source of complaint is not important, although Captain Jimenez will likely push you to find out who made the complaint. Reinforce the message that you will investigate issues that are brought to your attention without preconceived notions of guilt or innocence and even without knowing whether the issue has any validity.

4. What are two other alternatives you could use?

One alternative you might consider is observing a few shift changes for yourself. You may start at a distance to see whether Captain Jimenez's car enters the parking lot with only a few minutes to spare. If he is only infrequently rushing in at the last minute, you are not likely to observe his behavior firsthand. The failure to see the problem for yourself does bring into question the validity of the assertion made to you, because if Captain Jimenez is assigning his people to have things ready to go when he gets there, then perhaps the issue does not arise as frequently as they may want you to believe. For some people, one infraction may be sufficient cause for protest. If you do see Captain Jimenez entering the parking lot with only minutes to spare before shift change, be prepared to question him and hear his explanation. Again, he may have a perfectly good reason for the reporting time. If he tries to mislead you, however, you will have some firsthand observations to back you up.

A second alternative might be to ask the station officer working the previous shift what time he or she gets relieved by Captain Jimenez. If the previous shift's station officer is being kept at work longer than necessary due to another officer's personal reasons—that is, not detailed from another station or on a call—he or she will likely tell the truth about the situation. Because the officer may be hesitant to "tattle" on a fellow officer, ask open-ended questions: "Generally speaking, what time does Captain Jimenez relieve you?" and, "How consistent is that reporting time?"

5. What is your last-resort action?

A last-resort action might be to spend some time in that station at shift change. In addition to communicating that you are monitoring the situation, this action allows you to establish your expectation by directing Captain Jimenez in how you want him to display appropriate preparation and responsibility to his crew and the offgoing shift. This last-resort action requires significant scheduling adjustment, time commitment, and travel for you. For it to be effective, you must invest time and energy to be at that station at least 15 to 20 minutes before shift change takes place to observe the full effect. You will likely also be kept at the station following the shift change, as other members are likely to approach you to discuss other issues, ask questions, and so forth.

6. If you reach your last-resort action, what effect will you be dealing with two weeks after taking action?

There will likely be three effects if you implement the last-resort action. First, Captain Jimenez's behavior will improve due to your direction and oversight if, in fact, there was a problem. Second, the members of the company will appreciate that you intervened to stop Captain Jimenez's behavior; again, this effect is likely to occur only if Captain Jimenez was actually reporting for duty at the last minute as alleged. Third, your presence in the station may prompt rumors that you are keeping too close of an eye on, and perhaps are even distrustful of, your personnel. This misinterpretation of your behavior will likely lead to anger in some members, so you will be faced with the task of implementing some damage control through increased communication regarding the reasons for your presence. You may also need to look for other unrelated positive activities of the captain and crew while you are present for which you can publicly praise them.

16. Lieutenant Under the Influence: Reported by Direct-Report Fire Fighter

▸ NFPA 1021: 4.2.4, 4.2.5, 5.2.1

Lieutenant Jones has been detailed to your shift from another shift to serve as your company officer for Engine 5 until the regularly assigned Engine 5 lieutenant recovers from an injury. Lieutenant Jones has a reputation for working hard and playing hard. She has been with the department for more than 12 years and has not had any disciplinary issues of which you are aware. Lieutenant Jones is rumored to have been out partying the night before her current shift, based on Facebook posts. Some of the fire fighters who were at the party called and posted to fellow fire fighters who are now on duty at Station 5 that Lieutenant Jones was pictured and referenced in comments that placed her at the party well into the evening. She reports for duty, assigns the daily chores, and disappears by moving to a remote area within the station. Anytime other crew members approach the area where Lieutenant Jones is working, she moves to another remote location in the station—an office to catch up on paperwork or her bunk room to make her bed. A text message from an off-duty fire fighter to a fire fighter friend assigned to Engine 5 that morning mentioned the Facebook posts and observations at the party. One of the Engine 5 fire fighters decided to make a phone call to you, the battalion commander, to let you know about the Facebook posts. The fire fighter raises questions about Lieutenant Jones's pre-work activities and her continued movements away from the remainder of the Engine 5 crew personnel.

1. If you do nothing, what will happen?

Your failure to act will propagate rumors of an officer reporting for work under the influence and getting away with it. This perception jeopardizes Lieutenant Jones's credibility, your effectiveness as Lieutenant Jones's supervisor, and the department's mission.

2. What do you say or do?

First, your actions will hinge upon when you are made aware of this report. If the call reporting this issue comes within 1 or 2 hours after the shift starts, you will handle it differently than if the report is made 10, 12, or 24 hours into the shift. The Engine 5 fire fighter has made a phone call to let you know about the Facebook posts and his observations about Lieutenant Jones's continued movements away from the remainder of the Engine 5 crew personnel. If it is still early in the shift, such as within an hour or two of the shift starting, you have not only an obligation to investigate but also an opportunity to set the record straight by either dealing with an infraction if it exists or dispelling the rumor if the accusations are inaccurate.

Once you are aware of the situation, you will need to report to Lieutenant Jones's station to make firsthand observations. Speak directly with Lieutenant Jones to determine whether there is observable evidence that she is impaired. If you have reason to believe she could be impaired, based on odors, actions, or appearance, combined with the report you received, you will need to notify the appropriate authorities within your department and follow your department's procedures for reasonable suspicion drug/alcohol testing. The results and process will likely be clearly defined by your department's drug/alcohol policy.

If you do not observe any signs of impairment, evidence of impairment, or anything suspicious, you should consider notifying your supervising chief officer of the report you received, the findings from your investigation, and your intention to document the situation for your records and remain vigilant for any future activity. Your supervisor may direct you to take specific next steps in accordance with your department or city/district policy. This step is important because, regardless of which follow-up steps are taken, you have completed the investigation process and did not leave assumptions of guilt or innocence to chance. The actions you take after receiving the test results, positive or negative, will be identified by your department's policy—either a zero-tolerance policy or one involving a number of steps for disposition.

If you receive the report of the suspected issue after the fact—for example, if the phone call comes only after the Engine 5 fire fighter musters the courage to call in the afternoon or evening of the work shift—your ability to investigate the particulars is limited. As you investigate, you will need to use tact, diplomacy, and patience to gather information without communicating preconceived guilt. You will have limited access to factual data to dispel any accusation. You will be unable to make firsthand observations of any odors, speech patterns, or other symptoms that might indicate any possible level of impairment. Even the text message from the off-duty fire fighter that reported the Facebook post or the Facebook post itself could be deleted as time passes, which would make it difficult to assess all the related facts. Despite these potential obstacles, you still need to conduct an investigation to prevent a future occurrence as well as to prevent future frivolous and unfounded accusations against any department member. Your investigation indicates to Lieutenant Jones that you will not back away from following up on a sensitive issue. Your investigation also communicates to the fire fighters involved in reporting the issue that if they were making an attempt to tarnish or question the lieutenant's integrity or effectiveness, their misguided or even deliberately false accusations will be discovered and not tolerated.

3. With whom do you address the issue?

After such a report is received, the general order of communications in such a situation should be (1) investigate through observations; (2) discuss the matter with the accused officer in a private, one-on-one setting; (3) discuss the situation with your supervisor or chief officer as department policy dictates (this can vary depending on the size and composition of the department); and (4) carefully and guardedly gather information from the crew in the station affected so you have additional firsthand observations to draw upon. You should engage in these communications with the crew privately and individually to prevent a "group story" from developing. You must exercise extreme tact and diplomacy during these conversations, being careful not to ask leading questions or ask for confirmation about any specific details concerning Lieutenant Jones. Ask the station personnel about their observations, and let the fire fighters tell you what they know or do not know. For example, you might say, "Tell me about the morning routine this morning."

4. What are two other alternatives you could use?

One alternative might be asking the Engine Company 5 fire fighter who reported this issue to you to file the report in writing, complete with names of witnesses, times, and details. Creating such a written report will require the fire fighter to organize his thoughts, commit to the facts he is reporting, and be accountable to the credibility of the accusations. Should Lieutenant Jones dispute the information, the fire fighter will not be able to excuse himself from the allegation by stating that is not what he reported, that the source was someone else, or that his report was based on only rumors and the speculations of others. Nevertheless, be careful not to casually dismiss an accusation that is not accompanied by a written report. Although the lack of a written report could suggest that the initial verbal report is not credible, if the reporting fire fighter does not deem the issue serious enough to put the complaint in writing, do not ignore the issue completely just because a written report was not filed. If Lieutenant Jones reported for duty under the influence and you were informed of that fact, you are not absolved of the supervisory responsibility to investigate because the reporting party refused to document his or her observations. Failure to follow up on such a report may implicate you as negligent in your supervisory duties.

A second alternative might be requesting assistance from the local law enforcement agency, if you enjoy a good working relationship with them. If fire department relations with law enforcement personnel are strained or nonexistent, you may not find this option helpful or even possible. Many jurisdictions have law enforcement officers stop by the fire station throughout the day or night. It may be beneficial to enlist a law enforcement officer to stop by Lieutenant Jones's station and strike up a conversation with her. This option works only if the report comes to you on the same shift of the alleged infraction, rather than days later.

Take advantage of the investigative techniques and experience of law enforcement officers regarding the determination of public intoxication, disorderly conduct, and driving under the influence. Law enforcement officers must make these determinations all the time; take advantage of their greater expertise in this area. You may wish to accompany the police officers during the conversation and then seek their conclusions privately after the encounter to assess which steps you might need to take.

5. What is your last-resort action?

The last-resort action would be to send Lieutenant Jones for a reasonable-suspicion drug/alcohol test, even if that testing is performed against her will. A department policy would need to be in place that allows for such an action, and this action may require a chief officer's direction or presence. To send someone for such testing creates an operational void that must be accommodated, carries an accusatory tone, and may be perceived as a message of lack of trust for your people. Such an action, however, if allowable by policy, will provide conclusive data about whether an infraction has been committed.

6. If you reach your last-resort action, what effect will you be dealing with two weeks after taking action?

This situation is likely to create a climate of mistrust, regardless of the results of the drug/alcohol test. If the allegations are true, some will seek to identify the person who made the initial report. If the allegations are false, Lieutenant Jones and her advocates will likely be looking closely at the station personnel to see who tried to incriminate an officer. It also is possible that the fire fighters and some of the company officers may feel that you betrayed them by investigating the allegations instead of trusting the fire fighters and officers on the shift to know better than to commit such an infraction. In extreme cases, attorneys may become involved, with Lieutenant Jones claiming breach of her civil rights and protections. Things can escalate quickly.

As the shift commander, you have a duty to the public, the department, and your fire fighters to make sure you do not turn a blind eye to issues in need of your supervision. Fulfilling this responsibility is not comfortable, it is not easy, and it may even make you wish you were back on the apparatus dragging hose or pulling ceiling. Remember, however, that your ability to handle issues like these is why you are a shift officer—a supervisor of supervisors.

17. Lieutenant Under the Influence: Reported by Off-Duty Fire Fighter

▸ NFPA 1021: 4.2.4, 4.2.5, 5.2.1

You receive a phone call from Ryan, a fire fighter in the department assigned to the previous day's shift. When Ryan stopped by Station 5 earlier in the day to pick up some supplies from his locker, some crew members on duty mentioned to him that Lieutenant Jones, their officer, was rumored to have been out partying the night before reporting for duty. These rumors were based on Facebook posts and conversations with some of the fire fighters from other stations and shifts who were at the party. According to Ryan, the crew said that Lieutenant Jones reported for duty, assigned the daily chores, and disappeared to remote areas within the station. Any time other crew members approached Lieutenant Jones, she was reported to move to another remote location in the station. Ryan called you late in the afternoon because, upon further reflection throughout the day, he thought you should know about the potential issue.

1. If you do nothing, what will happen?

A significant element in this case is the reality that knowledge of this issue is spreading throughout the department across multiple shifts. Further, Ryan has made an allegation against not a fellow member of the department but an officer. It takes some courage to call an officer out to a battalion chief about such a serious allegation. If you do not address the issue, your failure to act jeopardizes the department's leadership credibility. Inaction communicates a message of indifference, possibly favoritism, and lack of respect for Ryan, who has brought the issue to your attention. All of these issues are unhealthy for the department, compromise your integrity as a leader, and, if the allegations are true, may potentially put the public at risk.

2. What do you say or do?

First, you must acknowledge the report and gather all available information from Ryan, including how he learned of the issue, when he learned of the issue, who else is able to provide facts or observations, and so on. You should then pay a visit to the station where the events occurred. If the incident Ryan reported is still ongoing, you should make this visit a high priority to see for yourself what is unfolding. If the report references a previous occurrence, then the visit may be important but not necessarily urgent.

At the station, gather the facts. Assess who knows what, what specifically they know or observed, and when and where it occurred. It is important to speak directly with Lieutenant Jones about the events that were reported. This must be a private conversation, and it requires you to be sensitive to the objective of gathering factual information without making accusations or conveying the message that she has already been found guilty.

Your department's policies will guide you in which communications are appropriate, which processes must be followed, and which testing might be needed. If your department does not have any policies on the subject, utilize your fire service network. Contact fellow officers you have met through conferences, seminars, academies, neighboring departments, or the Internet to determine other fire departments' policies on the subject; use those policies to guide your own process. Regardless of which actions you take, you should strive to preserve confidentiality, presume innocence until proven otherwise, protect the officer's individual rights and civil liberties against unlawful searches, and address the issue thoroughly so that nothing is left to chance or assumption. Through it all, you must communicate using your chain of command so your supervising officers and department administration are aware of the situation as it unfolds.

3. With whom do you address the issue?

Gather as much information from Ryan as possible, especially because the report in this scenario came from an off-duty member. Because Ryan is off duty, you will not regularly see or have access to him when you are on shift, and you will typically be away from work when he is on shift. Should additional information be needed from him as you investigate, he may or may not be available to take your call to provide additional details. Make an effort to gather as much information about the events while he is calling you with the initial report. After completing this step, the general order of communications should be (1) investigate through observations; (2) discuss the matter with Lieutenant Jones in a private, one-on-one setting; (3) discuss the situation with your supervisor or chief officer as department policy dictates (this can vary depending on the size and composition of the department); and (4) carefully and guardedly gather information from the crew in Station 5 so you have additional firsthand observations to draw upon in your investigation. You should engage in these communications with the Station 5 crew privately and individually to prevent a "group story" from developing. You must exercise extreme tact and diplomacy during these conversations, being careful not to ask leading questions or ask for confirmation about any specifics concerning Lieutenant Jones. Ask the fire fighters about their own observations, and let them tell you what they know or do not know.

4. What are two other alternatives you could use?

One alternative would be for you to contact your supervisor and request assistance from the department's (or municipality's) human resources department. A medium to large organization will most likely have such a department in place. This option may give the issue a higher profile, but you can benefit from the expertise and experience of someone who understands the finer legal points and knows the process to handle the issue well if you are able to provide the human resources staff with reliable initial information.

A second alternative might be to privately confront Lieutenant Jones directly and inform her of the report that she engaged in party activities prior to reporting for work that could jeopardize her ability to meet expected job performance levels. If you skip any discussion of her station movements of the morning, Lieutenant Jones will not be able to determine whether the report came from within the station or from an outside source. By taking the direct approach, you allow Lieutenant Jones to explain what occurred prior to her reporting to work and what has occurred since she came to work. Lieutenant Jones may explain everything satisfactorily and dispel any concerns directly. Alternatively, she may decide not to explain anything, placing the burden of further investigation upon you. The direct approach allows you to inform her that you are asking the questions about her activities prior to and during the start of the work shift for a specific reason. This option removes any ambiguity about the focus of your investigation, and it signals the level of seriousness of the report yet allows you to show support to the lieutenant by informing her you need to ask questions but you have no preconceived ideas regarding her guilt or innocence. Remember, unless your department has a policy in place prohibiting activity for a period of time prior to reporting for duty, you will have no jurisdiction over or responsibility for actions that occurred outside the workplace. Your responsibility to your department's members and the public is to ensure a safe, effective response to the public's emergency needs.

All departments typically share the expectation that members (both career and volunteer) and officers will conduct themselves in a professional manner at all times. Some departments may even have policies in place requiring all members to conduct themselves in a manner becoming to a member or officer at all times.

5. What is your last-resort action?

The last-resort action would be to send Lieutenant Jones for a reasonable suspicion drug/alcohol test, even if that testing occurs against her will. A department policy would need to be in place to allow for such an action, and such testing may require a chief officer's direction or presence. Sending someone for such testing creates an operational void that must be accommodated, sets an accusatory tone, and signals a lack of trust for your people. Nevertheless, such an action, if allowable under current policy, will provide conclusive data about whether an infraction has been committed.

6. If you reach your last-resort action, what effect will you be dealing with two weeks after taking action?

There is likely to be a climate of mistrust among department members, regardless of the results of the drug/alcohol test. If the allegations are true, some will seek to identify the person who made the initial report. The consequences of such behaviors on the part of a department officer may also create uncertainty and possible movements—promotions, demotions, or lateral movements—within leadership, all of which have an effect on the fire fighters. If the allegations are false, Lieutenant Jones and her advocates will want to determine who made the complaint. Because the issue was discovered by someone outside of Lieutenant Jones's company, her company may feel betrayed by another fire fighter who is not part of their company and harbor resentment and retaliatory feelings. Fire fighters, and possibly some of the company officers in other stations, may feel that you have betrayed them by investigating the allegations instead of trusting that Lieutenant Jones would not conduct herself in such an unbecoming manner.

18. Pornographic Material in the Station

▸ NFPA 1021: 4.2.5, 5.2.1

You stopped by Station 11 to deliver some requested supplies from headquarters, only to find that no one else is in the station. The company had been dispatched to a fire in a neighboring district, but it was reassigned while en route to stand by in the neighboring district's fire station until some of the initial response units were released from the fire scene. While walking through the station to drop off the supplies in the kitchen, you look across the room and notice a pornographic magazine lying on the recliner in the TV room.

1. If you do nothing, what will happen?

If someone from the public, a family member or child of a fire fighter working that shift, or a fellow crew member discovers the magazine in the TV room, its presence will become a problematic issue. As the shift supervisor, your responsibility is to support, lead, and protect the effectiveness and efficiency of the department's operations. If you do nothing when confronted with this scenario, you are missing an opportunity to prevent future trouble for the department and for your people. You also have a duty to the public. Laws place age restrictions on access to such adult material. In consideration of this fact, and in the context of the fire station being a public building and the fire department serving people of all ages, you have the duty to protect all of the public from exposure to the pornographic material. Your subordinate officers of the organization share that duty as well and should not allow this material to be present in the fire station.

Equally important, you must recognize how this material can contribute to creating an atmosphere of sexual harassment on an internal basis, among members of the department. Such material has zero value to public or the organization in terms of providing emergency services, response readiness, or public service.

2. What do you say or do?

First, remove the material from a location that is accessible to the public. You may place the magazine in the station officer's desk or a file cabinet in a location where it will not be stumbled upon by the public or another fire fighter. Check the magazine's cover to see if a subscriber's name is listed. Next you must communicate what you have done with the station officer so the magazine's owner cannot claim that you stole something that was rightfully his or hers. Finally, you will need to consult department policy to determine whether any regulations or departmental policies were disobeyed. Some departments have rules about what can and cannot be read, viewed, or listened to while on duty. As an officer, you should always make sure the officers under your command carry out their daily duties in accordance with department policies, in both emergency and nonemergency situations. This matter, if addressed by department policy, is no exception. If a policy exists, use it for guidance. If no such policy or guidance exists, then you must rely on the fact that the material provides no value in the department's service to the public, and because of the legal restrictions on access to such material, there is no rationale for it to be in the fire station at all.

Do not simply put the magazine in the trashcan and consider the matter resolved. The magazine belongs to neither you nor the department and is not yours to discard. Further, should someone from the public, a family member, or fellow fire fighter discover the discarded magazine, you will still have a problem on your hands.

3. With whom do you address the issue?

Communicate with the station officer. It is important to give your officers the chance to address matters themselves. If you address the issue on the officer's behalf, he or she will not receive the opportunity to learn how to handle this issue. This communication loop is fairly small: Meet with the company officer to communicate your expectation that he or she will address the situation, allow the officer to address the culprit (with you present at the company officer's discretion), and follow up with the officer to let him or her know you are available for ongoing support if needed. The loop becomes even smaller if the officer is the culprit: You deal with the company officer directly.

4. What are two other alternatives you could use?

You could address the crew later and present a hypothetical scenario in which you report the discovery of the magazine, explain where it was found, and ask the question, "How should I respond to the member of the public who stopped by the station when you were out earlier and found this as we strolled through the TV room?" Explain how one person's action can harm the entire crew, and even the entire department. Reassure the crew of your understanding that some members in the company, including the officer, may have not even known about the situation with the magazine; however, its discovery reveals just how quickly innocent crew members can be implicated by something about which they had no knowledge. Your unexpected observation of the material emphasizes the importance of everyone being watchful and protecting the integrity of the company and the department. Furthermore, once the public's perception is damaged or little eyes see something they should not have, that outcome cannot be undone.

Another alternative is to distribute a reminder memo to all personnel on your shift, noting that you found inappropriate material lying around Station 11 during a recent visit. Explain that rather than coming down hard and launching a full-scale investigation, you are using this event as an opportunity to remind everyone under your command of your expectations regarding keeping the station, and all areas within it, appropriate for the public, families, and children at all times. Use this incident as a springboard to reinforce existing policies or general expectations of conduct. If you opt to take this path, make it clear that the memo applies to all stations and not just Station 11, where you found the inappropriate material.

It is also important to communicate to the Station 11 company officer that you found the material in his or her station. The company officer needs to be aware of this fact so that he or she can monitor the situation. Ensure that the company officer does not feel as if he or she is under attack. Communicate your understanding that even though the magazine was found in his or her station on his or her shift, the owner of the magazine may not have been a member of that particular crew or even a member of the fire department. Encourage the company officer to fully investigate the matter, even if this investigation leads to personnel on the preceding or following shift (including volunteers).

5. What is your last-resort action?

If all other methods of station officer empowerment, common policy reminders, and self-policing actions fail and the problem persists, a last-resort action might be to conduct a surprise inspection of Station 11, accompanied by your company officer, to assess whether the station—including offices, TV rooms, kitchen areas, public reception areas, and restrooms—is clear of inappropriate materials. Any material found may be subject to disposal if advance notice has been given that such magazines are not to be present in the fire station. You may need to conduct these inspections periodically for some time to ensure you communicate the importance of the issue to all and to confirm that the personnel do not view it as a fleeting hot topic.

6. If you reach your last-resort action, what effect will you be dealing with two weeks after taking action?

This action requires increased time commitment by you as the shift battalion chief. Some personnel may feel that you are unjustly prioritizing the issue over staffing, equipping, training, responding to public calls for service, and long-term planning. Encourage your station officers to intervene and keep the situation from reaching the point where it requires you to conduct routine station inspections.

19. Citizen Complaint: Transport Delay

▸ NFPA 1021: 4.3.2, 5.2.1, 5.4.5

You receive an e-mail from the operations chief concerning a citizen complaint that alleges one of your ambulances, Ambulance 7, experienced navigation issues last week en route to the hospital, which resulted in delayed patient treatment. According to the operations chief, the patient's family has complained that the ambulance driver did not know where he was going and that as a result of the long travel time getting the patient to the hospital, the patient suffered adverse effects, including delayed surgical intervention and spread of infection. You have asked for a report from the dispatch center regarding Ambulance 7's incident times for that particular call. A quick review of the computer-aided dispatch times reveals that the transport took 10 minutes longer than is normal to reach the destination hospital.

1. If you do nothing, what will happen?

If you ignore this issue, you fail to complete an investigation task assigned to you by your supervising officer—and that is not acceptable. The alleged incident occurred on your shift, so you have a responsibility to investigate the situation and report back to the operations chief. Additionally, if you do nothing, you could be contributing to a future incident if one of your members is really unaware of how to reach a destination hospital.

2. What do you say or do?

First, relay to your operations chief that you are clear on the complaint and you will investigate it. Next, research the incident and determine whether the incident report provides information about the cause of the delay. After reviewing the incident report, you could investigate other objective data that may have influenced the transport time; for example, you might call the department of public works or highways to see whether any specific construction projects or road closures along the transport route might have slowed the ambulance's progress. One purpose in obtaining these kinds of highway data is to give the benefit of the doubt to your people. This step demonstrates that you have not assumed the driver was directly responsible for the delay and that you expect the delay to have been caused by some factor outside the control of your personnel. Finally, discuss the issue with the company officer supervising Ambulance 7 and possibly the Ambulance 7 crew to gather additional firsthand information.

3. With whom do you address the issue?

Discuss the data with someone from the highway department, the department of public works, or a similar department. This communication does not require you to go into detail about the specific incident, but you might let the respondent know you are reviewing response routes and times.

Your next conversation should be with the company officer of Ambulance 7. Ask open-ended questions about the date and time of the incident, and assess whether the company officer was aware of, or possibly even addressed, the delay with the Ambulance 7 crew before you became aware of the situation. The officer may have reviewed the crew's reports and identified the longer transport time based on a quality assurance review. If this happens, appreciate that the officer understands and takes his or her responsibilities seriously. If the officer is not aware of the situation, brief him or her on the information you have from the operations chief. You do not want to leave the officer in the dark about an issue with his or her people.

You will also need to speak with the crew to gain firsthand information about what happened on this run, but you may want to discuss with your company officer how that conversation should take place. Your officer may want you present at this discussion, or he or she may want to handle the initial discussion by him- or herself. You want to create and sustain a climate where the crew does not feel under attack but feels comfortable opening up about the details of the incident. If a driver error occurred, the Ambulance 7 crew will already be very aware of it.

4. What are two other alternatives you could use?

One alternative might be to investigate the driver's past transports to that specific hospital and compare the length of each transport. It will be helpful to know whether the delay was a driver anomaly. If it was, further investigation will allow you to learn the reason behind it.

A second alternative might be to use this event as a springboard to spot check a few other crews' transport times and assess whether you have discovered a need to provide some in-service training on hospital locations, driving routes, and alternate routes. This step might include a street map quiz for all drivers or focused attention on transport times for all crew members. Ensure that all reasonable steps have been taken to give the members proper training and continuing education regarding road closures and route changes. If you choose to implement this alternative, it will be crucial to focus on the mission—not the specific crew or response that brought this issue to your attention. However, you should begin this intervention with a review of the Ambulance 7 crew who was involved in the transport incident for which the complaint was received. This step allows for an accurate assessment of that particular incident to occur simultaneously with measures to improve performance. Other responses that day for Ambulance 7 or for ambulances stationed in that area that day would be the next likely subjects for study, which would contribute to both the complaint resolution and department training.

5. What is your last-resort action?

The last-resort action would be a form of disciplinary action. The level of discipline would be dictated by the facts you discover during the investigation. If your investigation reveals gross negligence or misconduct, severe disciplinary action for the driver may be warranted. This option is considered a last-resort action because it is unlikely that members of emergency services would intentionally delay a transport for care. Even so, you as a battalion chief must not dismiss this possibility during the investigation.

If you overreact by instituting a severe punishment such as a multiple-day suspension or proposed loss of driving assignment for what was an isolated, innocent mistake made by the driver, such as a missed turn at a particular street, you risk creating a climate in which your officers and their assigned personnel are afraid to make a wrong turn or a poor decision. Your expectation should be a top-notch performance and a quick recovery from any mistakes made, not necessarily perfection. At the same time, you do not want personnel to believe that a lack of training or poor performance or judgment is tolerated or excused. You also want your company officers to be aware they play an important role in ensuring the appropriate performance is attained, before the operations chief and the battalion chief bring poor performance to the company officer's attention.

6. If you reach your last-resort action, what effect will you be dealing with two weeks after taking action?

If discipline is warranted and carried out, you will probably face some divisiveness over the matter among the field personnel under your command. Some fire fighters will likely be on edge about making any mistakes. Other fire fighters who have not had navigation challenges or transport delays may voice their belief that the Ambulance 7 driver should have known better. Also, given the operations chief's involvement and awareness of the mistakes, the department administration may be dealing with further legal battles over the issue.

20. Disappearing Station Supplies

▸ NFPA 1021: 4.2.6, 5.2.1

At one of your fire stations, supplies seem to be disappearing. Because station supply order forms are submitted to you for processing and delivery of requested station supplies when you make rounds, you are aware that certain common supplies are being ordered almost weekly for a station that should have an ample supply. Requisitions for toilet paper, kitchen scrub pads, dish soap, shop rags, vehicle wax, and even copy paper now seem to be submitted every 9 to 12 days. This was not always the case; indeed, the requests have become noticeably more frequent within the past three months or so. At first, you dismissed the unusual volume of requests as an anomaly. When you decided these requests must be a carry-over from other shifts and considered denying the supply requisitions as duplicates, your company officer, Sergeant Lucas, told you that she does not know where things are going. It is essential for the members of your shift to have the necessary supplies to complete their assignments, and the sergeant needs your help.

1. If you do nothing, what will happen?

Regardless of why the supplies are needed, your company officer, Sergeant Lucas, has requested your help. If you do nothing, you will not fulfill your responsibility as a shift commander to support your personnel. Such a lack of support is likely to damage your reputation for good leadership not just in the eyes of Sergeant Lucas but in others' eyes as well. Additionally, if you do nothing, the root cause of the supply disappearance will go unchecked. If misappropriation, hoarding, or stashing is occurring, it will likely continue until it is addressed; it is your responsibility to do so.

2. What do you say or do?

First, gather information from Sergeant Lucas. Ask questions that she should be able to answer now or gather information to answer over time. Do not be quick to get to the bottom of the issue yourself; doing so would usurp Sergeant Lucas's authority to fix what is happening. By supporting her efforts to fix the supply problem on her own, you will encourage her professional growth.

Ask Sergeant Lucas the basic "who, what, when, where, why, and how" questions. Which supplies were ordered, in which quantities, when, and by whom? Is it possible that smaller quantities are being ordered and consumed more rapidly? How is the order being submitted? Word-of-mouth supply requisitions may indicate supplies were ordered but not purchased or delivered. Where are the supplies normally stored, and is there an inventory list logging supplies in and out? Are the station supplies secured in a locked room with limited access? If they are, who has a key or access to a key?

If your officer is unable to answer these questions, it is her responsibility to seek out the answers. She should also develop a system by which these answers can be obtained at any time, on any shift, for that station.

3. With whom do you address the issue?

Talk with Sergeant Lucas and encourage her to handle communications with her crew appropriately. You do not want to become the primary contact person for these issues. Managing such issues is the company officers' responsibility; as much as you may want to try to help, it can be all too easy to become overly involved in company issues until you are overwhelmed and increasingly ineffective in handling your own responsibilities.

You might consider mentioning the issue and the assignment to your peer shift commanders to keep them informed of station issues. All shifts need the supplies to complete their assignments, so a shortage would affect everyone at the station. Request that the shift commanders on the other shifts assess how frequently they receive these same supply requisitions, if any at all, or if their company officers in that station have experienced any shortages of needed supplies. Let them know you have assigned Sergeant Lucas to investigate the situation, so they will be informed about the problem. Be cautious in your timing and details so the other shifts' personnel do not perceive the message as a backhanded accusation.

4. What are two other alternatives you could use?

One alternative would be to inform your peer battalion chiefs of Sergeant Lucas's report, and ask them to share information with you about supply requests they have received or processed from that particular station. By tracking what was ordered and when, you may discover a pattern indicating which steps should be taken next. These steps could include holding conversations with specific officers, securing supplies, or even denying supply requests.

Another alternative might be to suggest that Sergeant Lucas implement an inventory system, if one is not already in place. Many manufacturing companies use a just-in-time inventory system that tracks the amount of time between when an order is placed and when the materials are delivered and ready for production. These companies maintain just enough inventory for use over the reorder/delivery window of time. Some times vary significantly depending on the item and the shipping origin. You might suggest that Sergeant Lucas research and implement a similar system for the station—one that will ensure an adequate supply is on hand and that includes a reorder trigger for supplies when the window for reorder-to-delivery time is reached.

Even if this inventory system option is implemented, you should still investigate further to determine what is happening to the supplies. It is not acceptable to simply order more supplies time after time without regard to the financial impacts on the department as a whole. You have responsibilities both to your personnel to support them in their assignments and to your organization to use its financial resources wisely and to seek out sources of waste or abuse and correct those problems. With deliberate effort, both aspects can be fulfilled. Size up the data and the circumstances, and develop your action plan—just like on any emergency.

5. What is your last-resort action?

The last-resort action would be for you to question the company officers in that station on the shift before yours, the officers in the shift after yours, and any volunteer officers who might be assigned to the station. Ask them the same questions you asked Sergeant Lucas. Approach them openly, and make every effort to avoid accusations of any wrongdoing—you simply want to gather information. This option is a last resort because it will require you to organize a meeting with the other shift officers. Your direct involvement will probably create some tension between the other officers and Sergeant Lucas. Your words may say you are not accusing anyone of anything but simply gathering information; however, the fact that you are a shift command officer who approaches the company officers from the other shifts may be perceived as furthering Sergeant Lucas's attempt to accuse another shift's fire fighters or officers of being responsible for the missing supplies. Your action may also create some tension with your peer shift command officers if you do not communicate about it in advance or if you have spoken to your peers and they believe the issue is not a big deal. When you address issues across all shifts, or with volunteers (if your department has them), you raise the profile of the incident.

6. If you reach your last-resort action, what effect will you be dealing with two weeks after taking action?

Ideally, you will have reduced consumption of supplies. Even if the cause was not fully identified, should any supplies be mishandled, the increased attention to this issue will raise its profile and will likely reduce its occurrence. It is equally possible that tension will increase among the shifts at that station owing to the other officers' or fire fighters' perceptions that you are leveling unspoken accusations at them. Inform Sergeant Lucas to be vigilant and stay ready to detect any possible changes in the level of cooperation or attitudes among the shift personnel in that station toward one another. Should a change be detected, do not dismiss it or delay investigating. Be ready to respectfully confront any attitude changes by talking about those issues with peer officers to suppress the volatility of the situation before personnel attitudes grow more negative.

21. Lieutenant Under the Influence: Not Formally Reported

▸ NFPA 1021: 4.2.4, 4.2.5, 5.2.1

You receive word through informal channels that one of your station officers is regularly noticed "popping pills" while on duty and at times acts "a little foggy." This information has come up in casual conversation around the stations while you make your daily rounds, but no one has directly told you who, when, what, or how often. You have repeatedly overheard your crew refer to an "Officer Foggy Lou," and today you heard a fire fighter say, "If anything's ailing you, the guy to see is the lieutenant. He has a prescription for everything."

1. If you do nothing, what will happen?

You have a potentially impaired member working in high-hazard, high-consequence events: If you do not address this issue quickly, both that individual and the department will potentially be at risk. You must act.

2. What do you say or do?

First, you must move past the inferences and gather substantive information. Clarify which officer is the subject of these comments, and gather information about the specific behaviors, actions, or inactions that have caused the fire fighters to claim the officer is not acting clearly. As the officer's immediate supervisor, you are the best person to investigate these allegations. You may be tempted to pass this problem up the chain of command because of the sensitive nature and potential consequences if the rumors are found to be true. If you pass this rumor up the chain for higher level officers to investigate, you miss the opportunity to do your job as the immediate supervisor. Involving higher level officers escalates the seriousness of the investigation, and their early involvement may even connote an assumption of guilt. You must do the preliminary work yourself.

After questioning your sources privately, follow leads to talk to whoever might provide additional facts. Be careful not to talk to too many other parties before meeting with the officer himself; information from two or three sources should provide a reasonable foundation for approaching the officer about the situation. When you approach the officer, explain that rumors have been casually shared in your presence and that you want to investigate the issue to clarify the facts and dispel any rumors or misperceptions. Let the officer know you are not accusing him. If he does not volunteer information, you may need to ask direct questions to discover the details: "Are you prescribed any medications that you take while on duty?" "Do these prescriptions have any warnings or side effects?" "Have you ever noticed any changes in your behavior after taking any of these medications?" The goal of this interview is not to pry into why the officer takes the medications or which medical issues he may be facing; rather, you are concerned with determining his ability to perform safely while on duty.

If your investigation reveals the presence of medication abuse of some type, communicate the facts to your supervisors, and refer to any policy or fringe benefit programs that may be available to the officer, such as employee assistance or rehabilitation programs. You will need to coordinate your next steps with the guidance for those programs or policies. If your department has no such policy or employee assistance counseling, then consider consulting with your insurance carrier or local medical practitioner for guidance. Do not assume that terminating the officer, implementing some other discipline, or ignoring the issue will fix the problem. You can also consult trade organizations such as the International Association of Fire Chiefs for assistance.

If the initial investigation reveals the rumors were unfounded and maliciously or recklessly spread by personnel, you may have to consider some disciplinary action, such as warnings or reprimands, for the individuals who spread the rumors.

3. With whom do you address the issue?

Speak with those who have directly observed the officer's behavior, and question only a few members to corroborate allegations before speaking directly with the officer. You may need to follow up with additional leads after having your discussion with the officer. In any event, out of respect for your officer, do not let him be the last to know about the investigation.

You may also have to communicate up the chain of command to your supervisors once you have facts to validate or debunk the rumors first brought to your attention. When you approach your supervisors, you should share accurate information with them rather than further spreading unfounded allegations through the organization.

4. What are two other alternatives you could use?

You could consult your local policies and contracts to assess whether random drug testing is allowable. You would still need to determine who the alleged officer is and whether his behavior is affected, so that a test could be performed to conclusively determine whether the officer is taking medications that are impairing his behavior. You must use caution if you choose this option. Moreover, your supervisors should not only be aware of the action but also concur with its implementation. You are entering into treacherous legal waters with this alternative, so communicate with your supervisors, and potentially legal counsel, before treading this path. On the bright side, this alternative can yield conclusive results and may be worth the effort to put the issue to bed once and for all.

A second alternative might be to identify the officer to which the rumors refer and deal directly with the officer first. Ask the officer to explain anything he feels might help clarify the origin of the allegations and any circumstances of which you would otherwise be unaware. Allow your officer to be the first to explain. If you take this approach, you must be prepared to weather a barrage of questions from the officer about the source of the rumors. If you have this information, you may not want to disclose it to prevent the officer from retaliating or behaving hostilely toward his crew.

5. What is your last-resort action?

After definitively identifying the subject of the rumors, your last-resort action might be mandating the officer's participation in an employee assistance program (EAP). This step is a last resort because you must have strong evidence to substantiate the need to send someone to the EAP. Further, when you or the organization mandates someone to seek help for dealing with a significant issue, animosity will develop if the individual did not seek or want help. You may be able to use this action for volunteer as well as career personnel. Volunteers have a relationship with the organization, even if they are not paid for their service. Many programs exist for volunteer fire fighters, and you may be able to access one of these programs for your volunteers. Study your particular employee assistance program language before you act, and know what you can and cannot mandate.

Finally, if you reach this last-resort action, do not go it alone. Communicate up the chain of command what you have discovered, which actions you have taken, and which actions you want to take, up to and including this last-resort action. Make sure your organization is informed and in support of the action before implementing it. A supervising chief officer may want to step into the process because of the potential volatility of this situation.

6. If you reach your last-resort action, what effect will you be dealing with two weeks after taking action?

If the rumors are true, many of your fire fighters will likely be relieved that a quiet hazard has been addressed and removed. You may also find that the officer rebels against the company, the shift, and you as the battalion chief. You may even face some legal challenges regarding the organization's ability to mandate entry into the EAP. Over time, however, you may find that the EAP has properly assisted the officer with his circumstances and that any observable abuses have disappeared.

Attitude Issues

22. Disgruntled Company Officer

▸ NFPA 1021: 5.2.1

Two weeks ago, Lieutenant Ash was moved to your shift from another shift. He has a reputation among his former co-workers as a complainer, and even some of your fire fighters are aware of his constant complaining because they have previously worked with him on overtime or through shift trades with other fire fighters. Your shift personnel have also previously heard Lieutenant Ash grumbling at shift change. He has a long-standing reputation for continually making comments around the station about his dislike for the department and departmental decisions and direction. Targets of his discontent include the people he is assigned to supervise; the equipment the department purchases; the design of the apparatus, from water tank size to hose capacity; and even the things that wear out around the station and need repair. Lieutenant Ash does not hesitate to express his displeasure to his crew or his comrades working on the shifts before and after his own shift.

You are now Lieutenant Ash's supervising battalion chief, and he continues to gripe to those around him in his new assignment. His fire fighters have informed you during your regular station visits that this officer is not a happy man, and his negativity is going to be a challenge to them. Even the personnel from Lieutenant Ash's old shift, when you encounter them, mention that they are glad he is no longer a part of their shift, because he brings everyone's attitude down. You are unaware of any previous actions or discussions with Lieutenant Ash instigated by his former commanding officers about his behavior. You recognize his behavior can be a drag on morale on your shift, and it is evident from fire fighters' comments that department morale has already been affected.

1. If you do nothing, what will happen?

If you do nothing, the negativity will spread. If Lieutenant Ash persists in being excessively negative, others around him may be persuaded to see the bad in circumstances more readily than the good simply because they hear a negative perspective vocalized so often. A fire fighter who was somewhat indifferent over time may become one who reacts negatively toward department decisions or other circumstances because that is the filter to which he or she has become accustomed. When the number of individuals with this negative perspective increases, it reinforces the unhealthy culture, and consequently morale erodes. In addition, it becomes more difficult for those personnel who tend to have a positive outlook to withstand the influence of the constant complaining and griping of those around them.

Of course, it is unrealistic to think that every person in the department will like everything about the department all the time. The issue here, however, is not just a disgruntled member but a vocal disgruntled officer, which magnifies the impact of the comments. You must act to preserve the morale and integrity of the organization. Not only is Lieutenant Ash openly commenting on his dislike of the department, but he is also questioning departmental decisions and direction. This opposition to the department's mission erodes confidence in department leadership and creates a divisive climate, which can be detrimental for the long-term health of the organization.

2. What do you say or do?

You must be deliberate, direct, discreet, and willing to truly listen and then ready to communicate clear expectations. Act deliberately and make time to immediately speak with Lieutenant Ash. Do not put this task off, do not pass it off to someone else, and do not ignore it. If your department is a volunteer organization, call and make arrangements to meet with Lieutenant Ash as soon as it is practical given both of your schedules.

During your meeting, approach the situation by making a direct request of Lieutenant Ash: "I understand you have been vocal about your dislike for the department and the way things are being handled. Tell me about it."

Exercise discretion by having this conversation in private. You do not want company members to overhear the conversation and risk having them become worked up about all the things allegedly "wrong" with the department. If you are the opposite gender of the company officer, find a location where you are within sight of others but not within earshot of others. Although you need to have a candid conversation, if the situation is already volatile, do not put either person in a compromising position should an accusation of inappropriate conduct come forward.

Be willing to truly listen. If you are direct and seek the root cause of Lieutenant Ash's displeasure, be prepared to hear your name mentioned; as the supervising officer, it is very possible that your decisions or actions may be the object of this officer's frustration. Should this happen, remain calm and hear your officer out—probe even deeper with questions so you have full understanding of where he is coming from. You do not have to explain, agree, disagree, or correct him at this point; you simply need to pay attention and listen.

As a supervising battalion commander, you have legitimate authority to communicate your work assignment expectations to your company officers. The first expectation to make clear with Lieutenant Ash is that a company officer has a responsibility to maintain the integrity of the organization. If either he or his fire fighters do not agree with departmental decisions or direction, there are appropriate methods to express that disagreement. Continually making critical comments fosters negativity, however, and is neither productive nor appropriate for a company officer. For these reasons, it is prohibited.

Explain that you expect Lieutenant Ash to bring his complaints, dislikes, and attitudes to the supervisor for resolution or direction rather than sharing them with his crew. Make it clear to him that should you, as his supervising officer, learn that he has disobeyed this edict, it will be grounds for a charge of insubordination and conduct unbecoming an officer.

3. With whom do you address the issue?

In addition to communicating with Lieutenant Ash, you should consider sharing with your supervising officer your knowledge of Lieutenant Ash's discontent with the department and let him or her know that you are actively addressing the situation. This notification serves two purposes. First, it communicates to your supervisor that you are handling your responsibilities to supervise company officers. Second, it allows you to make the first move should rumors or statements reach your chiefs indicating that Lieutenant Ash is a disgruntled employee. Instead of being obliged to investigate the situation, which would likely entail calling upon you to answer their questions, they will be aware that you are already working to resolve the issue. They may provide you with any details they have learned, but they should already be aware you are fulfilling your responsibilities to maintain the order within the shift.

4. What are two other alternatives you could use?

One alternative, albeit an option that is not as effective as the previous course of action that addresses the source, would be to simultaneously address the source and the affected company. Keep the discussion to the smallest group possible so you do not inadvertently spread the discontent to people on your shift who have been otherwise unaffected up to that point. This alternative deals with the root cause and the consequences at the same time. The dialogue would still need to be direct and expectations clear; however, it would be inappropriate to reprimand Lieutenant Ash in the presence of his crew. The tone should focus more on future expectations than on past actions.

A second alternative to consider would be to give Lieutenant Ash a writing assignment, instructing him to draft a comprehensive list of his perceived problems with the department and a list of the best practices to address each problem, including any benefits to the department. Instruct him to submit the list directly to you for discussion within the week. The writing assignment would allow you to address the issue immediately without having to be present immediately. It also allows Lieutenant Ash time to express his issues without getting worked up over them. Another important aspect of this alternative is that it requires the officer to look at the positive aspects of his job and the department. This option also provides you with material to use during the face-to-face discussion later—specifically, positive aspects to inject should Lieutenant Ash become overly negative during your meeting.

5. What is your last-resort action?

The last-resort action is to start the demotion or dismissal process. This alternative should be considered only after all other options have been exhausted. It is intended to prevent the harm to the organization that can occur if persistent negativity—specifically on the part of officers within the department—is allowed to continue. Before you demote or dismiss the officer, you must have communicated your clear expectations to Lieutenant Ash, and he must have failed to fulfill or operate within those expectations. This step is neither simple nor easy, but it does communicate the importance of maintaining good order within the department for the benefit of all personnel.

6. If you reach your last-resort action, what effect will you be dealing with two weeks after taking action?

Your workforce will likely be shocked at such extreme action being taken. Some personnel may be relieved, as some people within the organization probably did not agree with or appreciate the lieutenant's persistent negativity. Many of these fire fighters might not openly express their appreciation to you or the department for removing the disgruntled officer, but some—perhaps deep down—will be both appreciative and relieved to know their organization strives to maintain a positive and effective work atmosphere.

23. Highly Qualified Company Officer Passes up Promotion Opportunities

▸ NFPA 1021: 5.2.3, 6.2.4

Captain Salitto, a company officer in your department, handles almost every assigned task or issue well. Nevertheless, when promotion opportunities arise for a battalion chief or staff officer position, Captain Salitto consistently chooses not to seek promotion, despite encouragement from other officers and fire fighters. As a result, the candidates who are promoted within the department, although qualified, are all significantly less qualified than Captain Salitto.

1. If you do nothing, what will happen?

The strength of a department reflects the effectiveness of its company-level officers. A successful department must have a mix of top performers at all levels from chief to line fire fighters, but it is the company officer who unifies these members into a team. Company officers have the greatest opportunity to effect a positive direction, represent the organization to the fire fighters and the public, represent and communicate issues and ideas from the fire fighters up the organization's hierarchy, and establish the tone for daily operations. If you do not act, the department will continue to miss opportunities to build on its strengths. A climate in which candidates with mediocre qualifications receive positive reinforcement will spread throughout the organization, and there is a risk of normalizing mediocrity. For example, Captain Salitto is constantly pursuing continuing educational opportunities and higher education through advanced degrees, yet many of the candidates who choose to participate in the department's promotion processes have only the minimum requirements to participate. Indeed, for some candidates, those qualifications were achieved long ago, and they have shown very little ongoing commitment to continuing education. Many of the candidates ultimately selected for promotion may be driven by benefits such as pay raises, authority to assign station duties, and vacation priority rather than being motivated to develop the skills and abilities needed to move the department forward through finding ways to improve customer service and support development of younger department members. As a result, some of the line fire fighters who have skills and ambition will increasingly be supervised by these officers who carry out the basic functions of the job but are not preparing the future leadership of the department. As increasing numbers of these candidates for promotion are selected because of the smaller pool of applicants, the membership may start to believe that the department is seeking to promote officers who do the bare minimum—the basic job functions—and nothing more.

You may not have the ability to affect the outcome of the promotion processes or the final selections, but you cannot afford to ignore the developing situation and do nothing about encouraging personnel with the appropriate attitudes and qualifications to participate in these processes. As a shift commander, your organization and your people need you to engage the situation.

2. What do you say or do?

This is a communication issue. It is important for you to build on the positive observations of Captain Salitto's performance in his existing role. This situation can be a delicate one to address, because you do not want to erode the confidence in his fellow officers who have been promoted previously. In speaking with Captain Salitto, you must convey the observations that have persuaded you that Captain Salitto has the ability to contribute to the department beyond his current assignment.

Equally important is that you prepare yourself to listen intently to what this officer has to say both directly and indirectly about the promotion opportunities. As a shift command officer, you have the responsibility to analyze and interpret situational facts beyond the obvious or the superficial. You should be ready to empathize with personal situations, open to organizational criticisms, and willing to mine information discovered through previously unspoken conflicts—even about you or other supervising officers. You do not want to explain, defend, or push Captain Salitto into a particular action. For example, you might learn that he has not participated in previous promotion processes owing to his perception that the department needs to reconsider existing operational practices, schedules, or assignments. These observations should not be used as tools to manipulate or bribe Captain Salitto into seeking a promotion; however, you should be aware that there may be real issues that have been otherwise unnoticed by the entire department.

If Captain Salitto is ultimately promoted, consider which steps you have taken as a shift commander to effectively prepare another person to move into his vacancy. There are no guarantees that a particular person on your shift who is groomed for future leadership will be promoted in the future, but the preparations for that possibility should be considered and initiated while a talented company officer such as Captain Salitto is available in his existing position to develop line personnel working for and around him.

3. With whom do you address the issue?

You must speak directly and privately with Captain Salitto. This conversation could go in numerous directions—family issues, time constraints, continuing college education, side job responsibilities, positive or negative personal dynamics with current or future co-workers, and so on. The outcome of this first conversation with Captain Salitto will dictate which follow-up conversations should be pursued and which other personnel to involve.

4. What are two other alternatives you could use?

One alternative might be to seek counsel from a peer shift commander to make sure your observations are accurate regarding this company officer's capabilities. This discussion will give you a chance to identify any blind spots that might affect your perception of the officer before approaching Captain Salitto. Your peer may advise you of issues of which you were not aware. For example, perhaps Captain Salitto routinely leaves the station for the oncoming crew in less than a ready state, with housekeeping chores left incomplete or other station duties that are not readily known to the shift command officers. Conversely, if your observations that Captain Salitto is ready to increase the scope of his organizational responsibilities are shared by your fellow shift commanders, you might decide to arrange a meeting with Captain Salitto and the other shift commander present to discuss Captain Salitto's strengths and perceived opportunities. This meeting would serve to validate and reinforce Captain Salitto's positive work behaviors.

A word of caution: Captain Salitto may share limited feedback during the conversation. If he is facing challenges in his personal life, he may opt not to discuss these issues with anyone beyond his immediate supervisor. Be aware of this possibility so that you do not interpret limited feedback as an indication of any particular course of action. Regardless of Captain Salitto's decision about participating in the next promotion process, you will have affirmed his conduct as a company officer with this alternative.

A second alternative might be to develop a career map for Captain Salitto. After listening to what he has to say about why he has not participated in the promotion process in the past, you may seize the opportunity to discuss what he wants to achieve in the next 1, 3, 5, or even 10 years. This exercise would allow him to remain in complete control of the decision of whether he will participate in future promotion processes, while enabling you as the shift commander to fully develop this officer around his vision for his future. You may discover that the captain would become an excellent mentor for future officers, preparing them to take on the challenges he is either unable or unwilling to accept himself.

5. What is your last-resort action?

Your last-resort action requires your own acknowledgment that the decision about Captain Salitto's future is not up to you. Reaching this point means you have tried other options and have engaged to the fullest extent possible within your control and professional responsibility. Your last-resort action might be to simply ask whether Captain Salitto would be willing to fill in as an acting shift commander or other promotable position on a temporary basis. Your department may have such opportunities already in place, and this might be the best outcome available based on Captain Salitto's circumstances. If your department has not established an acting position, you may want to create one and select Captain Salitto to be the first officer to fill that role.

At this point, if you have clearly communicated your approval and your organizational needs to Captain Salitto, you have done all you can do. Now you must allow him to make his own choice, and you must be willing to support and encourage him in the role he chooses. You want to continue to seek the best outcome for the organization regardless of Captain Salitto's choice.

6. If you reach your last-resort action, what effect will you be dealing with two weeks after taking action?

There should be very little noticeable change. Those members who want to see Captain Salitto promoted will not be surprised if he does not participate. Those members wishing to be promoted themselves may be glad he decides not to participate. As the shift commander, your goal in managing the reaction is to prevent the normalization of mediocrity. Make it clear within those two weeks that opportunities to prepare for future promotions will be a priority for all members of your shift and your department. After you communicate this message to all of your shift personnel, follow through with your promises. Make sure the next promotion pool in your department is full of qualified, capable people, regardless of whether Captain Salitto participates in that process. Your department's future is relying on such preparations. Just like the initial response at a fire event, you cannot undo Captain Salitto's decision for his future. You must simply turn your attention to the remaining personnel and invest your energy into developing them for the future of the department.

24. Budget Cuts Reduce Staffing and Morale

▸ NFPA 1021: 4.2.6, 5.2.1, 5.4.6, 6.2.1, 6.4.6

Budget cuts have reduced your company staffing from four crew members to three, and your entire shift is upset. Several of the companies' members had worked together for many years, during which time they developed into effective response companies and became close friends. The remaining members are upset about having lost their colleagues. Judging from their expressions, kitchen table chatter, and lack of enthusiasm for their tasks, you know that some members of the companies are mad and discouraged. They are also concerned that with the reduced staffing, company assignments to complete tasks on emergency scenes will take longer, and they fear that they may even risk injury while trying to do so much in the first few critical minutes of the response. Sometimes, their discouragement boils over into anger at the circumstances that brought about the reduction in staffing. The staffing reductions were enacted only a month ago, so their full effects are only now becoming evident. You knew the staffing reductions were unpopular with the entire department even before they were implemented. The issue was bigger than just your shift and your shift management responsibilities, so you had not previously taken any actions. Now that morale and the personnel resources needed to conduct your shifts' activities are affected, however, you have a shift issue to manage as the shift's leader.

1. If you do nothing, what will happen?

If a situation on your shift is upsetting the entire crew, you have an obligation as a shift command officer to address the issue at hand. If you choose to do nothing, the shift's morale will continue to deteriorate. An issue of reduced staffing is a serious concern for fire fighters. If you do not act to address their concerns, other issues will become targets, and the level of discontent will rise. If you fail to apply good leadership skills to the shift through these changes, you will find the decreased morale makes your job more difficult, too. You might face the prospect of increased absenteeism, which in turn requires you to devote more staff time to covering the staffing shortages. You might note a decreased level of attention to detail or desire to fix what can be fixed, which in turn leads to increased equipment or apparatus down time, so that you have to make substitutions for these resources during incident action planning. You may feel that because the decision to reduce company staffing was beyond your control, you cannot really address the issue. As the leader, however, you *can* address your shift's attitudes and focus. Doing nothing is not an option.

2. What do you say or do?

The most professional steps you can take are to engage; to own the reality, instead of selling out the decision or the decision makers; to focus on what you can do instead of what you cannot do; to establish your clear expectation of a good attitude and appropriate focus while at the fire station; and to follow through. Some of these items require verbal communication, such as individual or company-wide crew meetings. Other items will require actions to back up the words and intolerance for slipping attitudes.

As a battalion chief, you have an obligation to do something about the shift's discontent—that is the "to engage" component of your response. Owning the reality instead of selling out the decision or the decision makers means that you accept the fact that this decision has been made and is not open to negotiation. When you communicate this fact to your shift personnel, do not feed the flames of discontent by blaming superiors, such as the chiefs, the council, the board, or city hall. As a shift command officer, you also cannot disassociate yourself from the department administration when times get tough. You can acknowledge the reality of a bleak financial situation or even acknowledge that the course of action is not ideal, but you must not blame others for the situation.

Too often, fire fighters have the misconception that administrative officers or city officials do not understand them and their priorities. When a shift command officer denounces the department's upper leadership or the community leadership, regardless of the issue under discussion, it strengthens this perception and further divides the organization. Reinforce the message that leadership generally wants what is best for the fire fighters and the community.

Focus on what you can do instead of what you cannot do. When you communicate with your personnel, you are likely to hear all the reasons for the shift's discontent. Assume the leadership role and redirect their focus to what they can accomplish. It is acceptable to acknowledge that these goals may take longer to achieve, but point out that they can still be accomplished well. When two fire fighters can divide the tasks of stretching the line and forcing the door, the attack is much quicker, but if only one fire fighter is available to do both tasks, the fire fighter is still able to perform both tasks safely and effectively. It may take the fire fighter longer to retrieve tools or carry hose, but he or she is still capable. Moreover, when the hose is in place and the door forced, the same work is accomplished. The fire fighters and company officers can drill more frequently to polish these skills and routines, and the fire fighters may have to brush up on some tactics in their down time around the station. Be clear, however, that all actions can be accomplished with three personnel on the apparatus instead of the more ideal four-member crew.

Establish your clear expectations about fire fighters' attitudes and focus while at the fire station. As shift commander, you have the legitimate authority to direct your personnel on how they should train and how they should act on an emergency scene and around the fire station. You can even lead them by identifying which work-related topics are unhealthy for the organization and instructing them that those topics are no longer to be discussed. Neither you nor the personnel on your shift are likely to have any control over the recent budget cuts and subsequent layoffs. Likewise, your complaints about the decision or the decision makers will not change the circumstances and are likely to distract the workforce from the objectives they are to accomplish. Follow through by enforcing your expectations for members to maintain a positive attitude and clear focus. Regardless of whether a member challenges you directly or makes an accidental misstep, it is important that you respond by demonstrating that you mean what you say.

Your people will be watching whether your expectations fade away over time or whether you follow through on your words by enforcing your expectations. This kind of enforcement could take the form of a private counseling talk or something more formal, depending on the severity of the member's action and the specific expectations that you established initially.

3. With whom do you address the issue?

Speak with your shift officers first. Remember, every time you deal directly with your shift fire fighters, you establish a direct link outside the departmental chain of command. Engaging them through casual social conversations, seeking general input or ideas, and communicating general information are acceptable. Conversely, if you counsel your fire fighters directly, you eliminate an opportunity for your company officers to perform their responsibilities. If you open the door to line personnel to discuss this issue, you must be prepared to leave it open when fire fighters approach you to vent directly, which can be a significant drain on your time. An exception is the case in which your company officer requests you to step in and assist with his or her own efforts in communicating with the line fire fighters. In such situations, you may deal with the crews directly, but you should simultaneously take the opportunity to encourage and grow your company officer so that he or she can deal successfully with such assignments without your assistance in the future.

Another person to consider consulting is your supervising officer. Let him or her know there is unrest on your shift about the staffing change, and inform this officer that you are actively addressing the situation. Supervising chiefs will undoubtedly appreciate the situational awareness communication and the knowledge that you are not letting the issue brew into a crisis unaddressed. You may receive an offer from your supervisor to assist you if needed, but reassuring the supervising officer that you are handling things will speak positively about you, your officers, and your shift.

4. What are two other alternatives you could use?

One alternative might be to focus on action rather than words. You may need your company officers to communicate to the shift personnel that your expectation is that they will adapt to and overcome the challenge rather than complain incessantly about the staffing reduction. Highlight actions that are within your personnel's control, such as company drills to assess performance and any response modifications that may be necessary to complete tasks with the available personnel. Reassure your shift that it is more important than ever that they wrap the hydrant, pull the hose line, and use the tools efficiently. Drills to hook to a hydrant and flow water are excellent training for a pump operator, requiring only time and a little clear space around a hydrant. Drills to deploy a hose line can be done on the apparatus pad or in an empty parking lot.

A second alternative might be assigning a research assignment. Do some preliminary research on your own to identify other fire departments that have historically operated with three-person companies. Caution: Do not look for guidance from departments that have recently experienced staffing reductions of their own; they will likely be going through the same anger and frustration, and you do not need to exacerbate these sentiments in either your department or theirs.

Consider assigning your company officers and their crews to contact a department that routinely operates with three-person companies and seek guidance on how it operates. This task shifts the crews' focus from the staff reduction in your department to development of a best practice of their own. Although there is no magic answer that will automatically make all of the fire fighters feel good about the change, this alternative is designed to make the best of a bad situation. No one wants to respond to emergencies with fewer personnel, but if all you have is three members for each company, find the best ways to use those three people. Other departments in the same situation may provide better input than if you try to figure out what gets cut from existing operations in your own department.

5. What is your last-resort action?

The last-resort action for this scenario would be to tighten up your expectations for your personnel. If you advise the personnel to focus on what they can do instead of what they cannot do, assign them to practice their skills, and seek out best practices for three-person company staffing, but the climate remains inappropriate and negative, then you may have to mandate a workplace attitude adjustment. This might require you to speak directly to the entire shift as a group if possible.

Inform the personnel that despite attempts to improve the climate through other alternatives, the workplace environment has not improved, and the fact remains that staffing will not change. Inform your personnel that they have an important choice to make: They can make the best of the three-person staffing, be thankful they are still members in the department that operates the three-person companies, and stop complaining about the issue immediately, or they can choose to leave this department and seek opportunities with another agency that provides the four-person company staffing they desire. If they make the former choice, they can hope that future changes will restore additional staffing; however, the topic is not currently open for discussion at the fire station. Conclude with a consequence statement that reinforces the message that each member has the ability to make the choice for himself or herself, but if they choose to stay and continue to complain or contribute any negativity to the existing operation, you will be forced to deal with a failure to follow lawful orders.

6. If you reach your last-resort action, what effect will you be dealing with two weeks after taking action?

This alternative is the last-resort action because it forces your personnel to make a direct and uncompromising choice. When you are no longer able to find room to compromise, you lose some of the tools in your leadership toolbox. This last-resort action means you no longer have the time or the attitudinal atmosphere that would permit personnel to bring up the topic, even on rare occasions. If you issue an order stating the topic is not to be discussed at the fire station, then any subsequent mention of the topic at the fire station must result in some consequence. The attachment of an enforcement action to your edict ensures the integrity of the directive and compels members to progress toward achieving a more positive work climate. If you select this option, the grumbling will likely continue for a time, but personnel will be more careful about what they say and where they say it. You will face a challenge, however, when one of your personnel inevitably lets a comment slip in your own or an officer's presence. You do not want to dismiss anyone over one comment, but you need to be prepared to take some sort of action in this circumstance.

Your supervisors may not like this rigid approach, but they may be pleased that you were willing to take the difficult route to achieve a better organizational climate when other alternatives did not succeed.

Realistically, this last-resort action may cause you to develop a reputation as a dictator, and some may see this as an opportunity to challenge you in the future. In summary, exhaust all other alternatives before taking this hard-line approach.

25. Company Dissatisfied with Design of New Apparatus

▸ NFPA 1021: 5.2.1, 5.4.6

A new piece of fire apparatus is assigned to one of your stations, Station 14, and your Engine 14 crew members do not like the newly received fire engine's compartment design. They think the storage compartment located inside the cab is both too big, reducing the room available for the crew, and poorly designed, because the equipment intended to go inside that compartment could just as easily be stored in other places. You are aware that Engine 14's crew is unhappy with the apparatus design: As you have visited the other stations under your command, you have heard fire fighters who have occasionally been detailed to Station 14 to cover an absence talk about the complaints Engine 14's crew have made to the Engine 14 lieutenant. This is the first brand-new apparatus the station has received in years; Station 14 has historically received apparatus transferred in from one of the busier stations within the department when those stations received a new apparatus.

1. If you do nothing, what will happen?

In some ways, it might be appropriate for you to do nothing. Your involvement at a shift command level will circumvent the Engine 14 lieutenant's role, and it will not change the fact that the Engine 14 apparatus is new and in good working order. However, if you do nothing, you risk two consequences. First, a negative culture will begin to develop and take hold within the membership—one that focuses on problems rather than on the accomplishment of having acquired a new apparatus for reliable use. Second, erosion of the respect and effectiveness of the Engine 14 lieutenant may occur. If the Engine 14 crew members have made their dislike of the apparatus well known to the company officer, and if they continue to spread their discontent to others in the department, you have a developing problem on your shift far greater than the issue of equipment placement. Because these secondary consequences will affect both the shift and the shift leadership, you will need to become involved in some way to preserve the positive organizational culture and leadership effectiveness—not necessarily to solve the question of which equipment goes where.

2. What do you say or do?

First, gather accurate information. Find out what the specific dislikes are and what the Engine 14 lieutenant has said or which actions she has taken prior to you learning of the issue. Assess whether your supervisor will allow you or your company officer any latitude to make changes to the apparatus design if needed. Remember that because other shifts would be also affected by any such changes, a final decision may come from a higher level in the organization for the sake of department-wide consistency.

3. With whom do you address the issue?

After documenting what you have heard about the Engine 14 fire fighters' complaints, meet with the Engine 14 lieutenant to privately discuss the situation with her. Let her know what you have heard and when this information came to you. This meeting allows you the opportunity to assess whether she has already discussed the matter with her crew. If she has, you have a new attitude issue to address: Your officer's subordinates are undermining her authority by escalating their complaint to a higher ranking officer in hopes of bringing about the desired change. In this instance, consequences may range from verbal counseling for the crew to charges of insubordination, depending on exactly what the Engine 14 lieutenant told her crew in response to their equipment complaint.

In this scenario, it is also important to identify those parties with whom you should *not* speak. Do not discuss the matter with the lieutenant's crew directly. A strong desire for results may tempt you to take direct action. If you engage directly with the Engine 14 crew, however, you open the door to having other issues dropped at your doorstep through indirect communication, with personnel expecting you to become involved or even reverse a company officer's decision. Allow your company officer to do her job and support her efforts, refraining from jumping into the fray unless she asks you to become involved.

If you learn the Engine 14 lieutenant has not taken any actions or made any decisions, inform her about the issues of which you have been made aware, and ask her to develop an action plan and report back to you.

4. What are two other alternatives you could use?

One alternative might be to informally ask your peer-level shift command officers if they have held any discussions with their officers about the new Engine 14 equipment placement. This inquiry could shed light on whether your company has been complaining directly to the other shifts, as well as whether the other shifts have identified the same or similar issues independently. If the other shifts desire a change as well, perhaps the company officers might hold discussions among the Engine 14 crews to address alternatives.

If you have already asked your supervisor whether you or the company officer has any latitude to make changes and the answer was "no," then you must respect that decision, even if all of the shifts want to make changes. Additionally, you must present that decision to your officers and crews as if it was your own. This step is important because your supervising officers have made a decision and communicated it to you that changes are not to occur to the Engine 14 equipment placement. Regardless of why that decision was made, you are now responsible for carrying out the directive because it is your duty to obey lawful orders you receive. This adherence to the higher ranking officers' decisions preserves the integrity of the chain of command within the organization and keeps communications clear and consistent.

A second alternative could include initially meeting with your Engine 14 lieutenant to determine whether she has already addressed this issue and directing her to assign her crew members to make a full written proposal seeking the change. The members should present this report to the Engine 14 lieutenant and, if it has merit, the report could subsequently be sent up the chain of command. The report should identify any advantages and disadvantages of the current equipment locations and the proposed location changes, as well as full cost disclosures and any out-of-service time required for the apparatus as a result of those changes. The order for the report is intended not to quell the verbal complaints but rather to encourage a culture of fully researching issues and presenting a finished product in the form of a polished written report rather than simply complaining verbally about problematic issues. This alternative may create some animosity if the crew perceives it as busy work or punishment for speaking out. The Engine 14 lieutenant will have to be prepared for this outcome and maintain a constructive tone within her crew.

5. What is your last-resort action?

The last-resort action will likely include a final decision to be communicated from you through the company officer that the decision on equipment placement has been made and is no longer open to suggestion or up for discussion. Opportunities for input have been extended and considered, but as with all issues with variable outcomes, when a decision has been reached, it is time to abide by that decision and move forward. Communicate that anyone heard debating, discussing, complaining, or reviving discontent will be considered as disobeying a lawful order or directive issued by the company officer.

6. If you reach your last-resort action, what effect will you be dealing with two weeks after taking action?

After you take this last-resort action, some personnel may feel as if they have been censored. Those fire fighters may try to push the boundaries of what is allowable, both with their company officer and with you as the battalion commander. You and the Engine 14 lieutenant must be prepared to face these challenges and stand your ground for enforcement. Here the greater issue of complying with an officer's directive drives the enforcement; the specific apparatus equipment placement becomes a secondary issue.

If a clear message is sent that the topic is no longer open to discussion, some personnel may move on to the next topic of controversy. Be prepared for personnel to occasionally attempt to move equipment as time goes by. If your fire fighters truly believe that the location changes are an improvement, they may choose to move equipment when they are working and hope it is not reported. Again, in such cases be ready to support the Engine 14 lieutenant as she enforces the directive for the sake of maintaining good order within the department. As the shift commander, you may support your company officer by reminding your crews that the true value of effective emergency response is not where equipment is located or even which equipment is available but rather which fire fighters use it—and that important feature has not changed.

26. Inappropriate Comments About a Patient During a Response

▸ NFPA 1021: 4.6.3, 5.2.1

You respond as the command officer to a motor vehicle collision with possible entrapment. Engine 2, Rescue 7, Ambulance 2, Ambulance 4, and law enforcement personnel are on scene when you arrive. The female driver of one of the vehicles is still inside the vehicle and is noticeably impaired. She smells strongly of alcohol, and empty beer bottles are visible in the vehicle. Two female passengers from the same vehicle are being packaged for transport with serious but non-life-threatening injuries, including trauma to extremities and lacerations. These passengers also smell of alcohol and are crying excessively. As the driver is being extricated, you hear Dave, one of the members from Rescue 7, comment to Mary, a fire fighter from Engine 2, "It's gals like this that keep us in business." There are spectators around the crash scene, and you are unsure if anyone has heard the comment. Nevertheless, it was spoken loudly enough for you to hear it, and some of the other fire fighters working around the car look up at Dave in response to his comment. Mary just makes eye contact with Dave, does not comment, and keeps working.

1. If you do nothing, what will happen?

If the patients do not file a complaint, you may think no action is needed. However, lack of a formal complaint does not make the comment appropriate or inaction on your part acceptable. In fact, just the opposite is true: Inaction may reinforce inappropriate behavior, which can impact the organizational culture.

As battalion chief, you are aware of what was said and of when and where it was said, and you have a duty to manage risks of which you are aware. As a battalion officer, you must identify and resolve issues before they come to you via public complaints. If you do nothing in this scenario, there may initially be no consequences. When this incident or subsequent occurrences are discovered, however, you will find yourself dealing with a much more complicated issue.

2. What do you say or do?

In this scenario, you are present and within earshot of the comment while the extrication is under way. It is appropriate to directly, sternly, and professionally redirect Dave's focus the moment you hear the comment. You could use a simple statement to do so: "Enough of that. Focus your actions and attention on the task of the extrication!" This rebuke sends a clear and quick message about the appropriate behavior. You do not want to draw unnecessary attention to the statement by anyone who has not already heard it. You also want to reassure the bystanders and spectators who have gathered around the crash scene, many of whom could easily have heard the comment, or the patient who may have heard the comment, that this statement is not acceptable by correcting it as soon as it is uttered.

After the call is over and everyone is back at the station, follow up with the Rescue 7 company officer and Dave, the fire fighter. Convey the importance of completing emergency tasks in accordance with the standards set for public safety services: to perform duties with respect to the dignity of all people, regardless of background or circumstances. Fire fighters must carry out their tasks fairly, impartially, and expediently to the best of their ability. Judgment statements during an emergency task are ineffective and counter to the oath of public service. Inform Dave of the possible legal consequences of making insensitive remarks during a response. The outcome you seek is, at minimum, for Dave to cease making further inappropriate comments and, ideally, to foster a more compassionate perspective in him.

3. With whom do you address the issue?

This scenario calls for a jump in the chain of command. Because you are present when the comment was made and have firsthand knowledge of what was said, you have the ability to deal swiftly with the situation yourself. Address Dave directly to refocus his attention on the vehicle extrication. It is important for everyone in the chain of command of any organization to recognize that as you move up the chain of command through your own promotion, you become responsible for both your own and your subordinate personnel's performance. As the battalion chief, you are ultimately responsible for the fire fighters, the company officer, and yourself. If a fire fighter steps out of line in your presence, you have the legitimate authority to swiftly correct this action to ensure the effectiveness of this specific response and for the overall good of the department.

You should also follow up with the Rescue 7 company officer after the call to inform her of what you observed, which actions you took, and which expectations you have for her, if any, regarding future counseling of the Rescue 7 fire fighter. Note that counseling does not necessarily mean a "write-up"; it can be mean a dialogue directed toward improving future performance. We often use "counseling" as a synonym for "disciplinary action," but the term does not have to carry this meaning.

4. What are two other alternatives you could use?

One alternative might be first redirecting Dave to focus on the extrication task and later discussing the ramifications of such comments back at the station by presenting Dave with some hypothetical scenarios. Privately meet with the Rescue 7 company officer and Dave. Ask Dave, "What would you say if I told you that one of the patients you commented about back on the emergency scene was my niece?" Give the fire fighter a chance to respond to you and to his company officer; he may feel uneasy in the moment. Then ask him, "Would your response change if I told you the patient was not my niece but rather the daughter of a prominent trial attorney? What if the patient was a prominent trial lawyer who likes to challenge city departments and has been known to file complaints with the fire chief in the past?" These questions may prompt the fire fighter to consider the ramifications of his words. If he does not see the connection or does not seem to care, explain that it is your responsibility to the department as a shift command officer both to ensure appropriate services are provided to the public and to minimize risks to the department. Comments such as the one Dave made at the scene are insensitive and inappropriate, create organizational risks, and cannot be tolerated.

A second alternative might include a written assignment. In a private meeting back at the station after the call, instruct Dave to write a report identifying, with quotes, all the references in the department standard operating guidelines and regulations that indicate his comment, "It's gals like this that keep us in business," is appropriate when providing services to the public. Then instruct him to identify all the standard operating guidelines and regulations that indicate the comment is inappropriate and quote the consequences identified in those department standards. Direct him to type and submit the report to the Rescue 7 company officer, who will then forward it you as the battalion chief. Explain that any future action will hinge upon the thoroughness of the report. This alternative clearly identifies the behavior as unacceptable and places the responsibility for meeting the department standards on such behavior on the person responsible.

5. What is your last-resort action?

The last-resort action would be to initiate formal disciplinary proceedings. Due to the unacceptable nature of the comment, you may believe a disciplinary proceeding is merited right away rather than as a last resort. The issue here is less than clear-cut, however. Although the comment was certainly unacceptable and should be addressed, the fact that it was a comment, rather than negligence to act or an improper action, might lead some to argue that the organization cannot regulate a personal opinion. Dave could argue that he continued to perform all of his assigned duties during the extrication, that he provided appropriate care to those in need, and that no regulation prohibits him from observing an emergency scene and forming a personal opinion. Disciplinary action could be disputed by Dave, and possibly any labor representation on his behalf, based on the defense that any disciplinary action for unbecoming conduct would be too extreme for merely an expression of his personal opinion. Although the statement could be construed as representative of the department, it is still an individual comment. The consequences of the fire fighter's remark might be very harmful to the organization, but a quest for his formal discipline for a first-time offense will be difficult to maintain in the face of scrutiny from a labor organization, a disciplinary appeal board, department members who sympathize with Dave, and possibly even segments of the public who believe the department over-reached its authority to regulate personal comments.

Although a formal proceeding would send a strong signal about what is and is not appropriate behavior, it would also draw public attention. At this point, if a formal complaint has not been filed or if others did not notice the comment, you want to keep the impact of Dave's comment as limited as possible while you counsel him to prevent any future occurrences. The decision to initiate formal disciplinary action absent a triggering event such as a public complaint will likely be vigorously contested by Dave or others who feel personal comments made on work time are not subject to employer controls. This challenge may easily evolve into a more public debate within the department or among the public at large. Nevertheless, even if neither the public bystanders nor the patients treated during that response register complaints, it would still be appropriate and necessary to conduct a counseling session with Dave to teach him the consequences of his statements and the adverse impact on the organization. He needs to be aware of the infraction, but you want to address it effectively with as little fanfare as possible. Disciplinary action will certainly publicize the issue to a larger audience.

If this is not the first offense, the situation changes. A first offense provides an opportunity to clarify expectations regarding personal comments and behaviors for which disciplinary action is merited. If instruction or counseling was provided after a prior offense and the behavior occurred again, it is necessary to create a consequence strong enough to communicate—without any ambiguity—that such comments are not acceptable and will not be tolerated. If a prior event occurred and the fire fighter received advance warning about avoiding any future events but failed to heed that warning, significant corrective action is now warranted.

6. If you reach your last-resort action, what effect will you be dealing with two weeks after taking action?

There will likely be challenges to the appropriateness of the action. The fire fighter's co-workers and possibly labor representative may challenge any punitive discipline and assert violations of individual freedoms of speech. You do not want to engage in a war of words over what is right and wrong to say. This last-resort action will create battle lines within the department—those in support of and those against it—and will possibly expose Dave's comments to the public or the media, which will magnify their ramifications. Consider other alternatives first. To use a size-up analogy, just as you would not use the deluge gun on a wisp of smoke from a trashcan fire, so you do not want to throw the big gun of disciplinary action at this personnel issue despite the potential significant consequences. The trashcan fire could become a raging inferno if not addressed quickly and appropriately, however; thus, you should size up this personnel issue quickly and appropriately and take swift action to prevent it from evolving into a full-blown crisis.

SECTION 3

Beyond the Battalion

Introduction

As you climb the organizational chain of command, you will find some of the same issues take on different perspectives. Up to this point you have had the opportunity to learn how to size up situations from the company officer perspective and the shift commander– or battalion-level perspective. You can see the linkage between each level of officer handling some of the same issues differently because there are different stakeholders and wider impacts to be considered.

Operations Chief

When you move to the operations chief level of the organization you may be referred to as assistant chief, deputy chief, division chief, or other similar titles. Regardless of specific title, the position is ultimately responsible for the entire internal operation of the department. This might include all shifts of career and volunteer personnel or even multiple types of services within the department. People in these positions still have to deal with and consider outward perspectives, but the majority of their focus is upon the consistent delivery of effective services to the public. It becomes increasingly difficult to stay immersed in all the functional details of the organization because the volume of various issues and responsibilities increases significantly.

Fire Chief

Just as a company or battalion officer might focus on the smoke, building, fire protection features, and so on when sizing up a street-level fire response, the operations chief may be sizing up staffing requirements for the remainder of the shift or into the next shift. The fire chief may be sizing up the fire response from a communications perspective with elected officials or the media. Maybe considerations of mutual aid requests for back filling stations of committed companies would be discussed by the operations chief or fire chief. These are important size-up analyses but have little to do with strategy or tactics at the fire itself.

Rather than attempt to identify the many scenarios individuals in these positions deal with regularly, as that would be another text altogether, there are a few select issues identified in the operations chief and the fire chief sections that carry on the previous scenarios' thread of commonality. The issues here are some of the same issues discussed earlier, but you can size up how they might be addressed at this level of the department. You will find operations chiefs and fire chiefs must consider a number of internal and external factors for the department if the issue has reached their desk.

As you continue the station issue size-up exercises, you are now approaching them from a different perspective. Do you find yourself looking at the actions and decisions differently? If you do, stop to consider how you increase your personal effectiveness for your organization if you size up an issue from all perspectives before you choose a course of action.

Chief Elected Official

In this section, the same scenarios are used for elected officials as for fire chiefs. The primary difference is for you to figuratively walk in the shoes of the city manager, the mayor, or the board chairperson. You will still implement your answers as the fire chief, but consider the questions as if you were the chief elected official. Effective chief-level officers must recognize the perspectives of those for whom they work.

If you are reviewing these scenarios for future advancement, this is the very real reminder that *everyone* has a boss, even the fire chief. In fact, many fire chiefs are appointed or elected by their bosses, and it is important to recognize you serve at the pleasure of that boss. Therefore, buckle up and let's take a ride into the world of the fire chief from a chief elected official's perspective.

CHAPTER 5

Operations Chief Scenarios

1. Career Members Question Volunteer Members' Skills

▸ NFPA 1021: 5.2.1, 5.4.6, 6.2.1, 6.2.7, 6.4.6

You are the operations chief of a combination fire department. Keeping career and volunteer members on the same page is a constant goal. A group of your career membership thinks that most of the volunteer members do not know the basic skills of being a fire fighter, such as breathing apparatus emergency procedures, hose stream selection and safe use, and even fire behavior recognition—all of which are necessary to perform successfully as a fire fighter. There is a growing sentiment among career members across all shifts that the volunteer members pose serious response capability risks and that more full-time staff are needed to fulfill emergency response roles. Recently, a few battalion chiefs have casually shared that career and volunteer members are speaking freely and frequently about the issue, which is beginning to polarize the department.

Today a vocal career fire fighter visits you in your office. He claims to represent the career fire fighter who first complained against a volunteer member and other career fire fighters who wish to remain anonymous but share similar concerns. He comes to you directly while off duty because the issue does not involve his company officer or battalion chief. He explains that the complaints began when the career member on the shift before his raised concerns about a volunteer fire fighter's skill deficiency with her company officer and later escalated the issue to her battalion chief when no action was taken. He emphasizes that various members have raised similar concerns numerous times with company and shift officers but that nothing has been done about it, and the ill-prepared volunteer members are still responding to calls.

1. **If you do nothing, what will happen?**
2. **What do you say or do?**
3. **To whom do you say it?**
4. **What are two other alternatives you could use?**
5. **What, if any, legal issues should you consider?**

6. **What is your time frame for addressing this issue?**
7. **What is your last-resort action?**
8. **If you reach your last-resort action, what effect will you be dealing with two weeks after taking action?**

2. Fire Fighter Fraternizes with City Manager

▸ NFPA 1021: 5.2.1, 5.4.6, 6.2.1, 6.2.7, 6.4.6

Your battalion chief stops by your office and asks for a moment to brief you on a situation that his station officer shared with him last shift: During routine physical fitness training, a fire fighter and the newly elected labor representative, Kevin, mentioned to the station officer that he has become increasingly friendly with the city manager. The lieutenant did not specifically know what to do or say so he mentioned it to the battalion chief to keep him informed. Your battalion chief tells you that he, too, is unsure of what to think of the situation but thought you should be aware of it.

Your battalion chief explains that Kevin did not previously know the city manager but that since Kevin's election, the city manager has extended dinner invitations to him and given him tickets to sporting events. In some instances, the city manager and Kevin have been the only city personnel in attendance. The city manager's family and Kevin's family have also attended multiple family picnics together. From what Kevin shared with his station officer, your battalion chief knows that fire department business sometimes enters Kevin's off-duty conversations with the city manager.

1. **If you do nothing, what will happen?**
2. **What do you say or do?**
3. **To whom do you say it?**
4. **What are two other alternatives you could use?**
5. **What, if any, legal issues should you consider?**
6. **What is your time frame for addressing this issue?**
7. **What is your last-resort action?**
8. **If you reach your last-resort action, what effect will you be dealing with two weeks after taking action?**

3. Citizen Complaint: Transport Delay

▸ NFPA 1021: 5.2.1, 5.4.6, 6.2.1, 6.2.7, 6.4.6

You receive an email from a citizen alleging that one of your ambulances, Ambulance 7, experienced navigation issues en route to the hospital, which resulted in delayed patient treatment. The email is from the patient's family and states that the ambulance driver did not know where he was going and that, as a result of the long travel time to the hospital, the patient suffered adverse effects, including delayed surgical intervention and spread of infection.

You have asked for a report from the dispatch center regarding the Ambulance 7 incident times for that particular call. A quick review of the computer-aided dispatch times revealed there was a 10-minute longer transport time than normal for the destination hospital.

FIGURE 5-1 A scathing complaint email from a citizen can reprioritize your day or even your week.

Courtesy of Jeff Barlow.

1. **If you do nothing, what will happen?**
2. **What do you say or do?**
3. **To whom do you say it?**
4. **What are two other alternatives you could use?**
5. **What, if any, legal issues should you consider?**
6. **What is your time frame for addressing this issue?**
7. **What is your last-resort action?**
8. **If you reach your last-resort action, what effect will you be dealing with two weeks after taking action?**

CHAPTER 6

Operations Chief Discussion

1. Career Members Question Volunteer Members' Skills

▸ NFPA 1021: 5.2.1, 5.4.6, 6.2.1, 6.2.7, 6.4.6

You are the operations chief of a combination fire department. Keeping career and volunteer members on the same page is a constant goal. A group of your career membership thinks that most of the volunteer members do not know the basic skills of being a fire fighter, such as breathing apparatus emergency procedures, hose stream selection and safe use, and even fire behavior recognition—all of which are necessary to perform successfully as a fire fighter. There is a growing sentiment among career members across all shifts that the volunteer members pose serious response capability risks and that more full-time staff are needed to fulfill emergency response roles. Recently, a few battalion chiefs have casually shared that career and volunteer members are speaking freely and frequently about the issue, which is beginning to polarize the department.

Today a vocal career fire fighter visits you in your office. He claims to represent the career fire fighter who first complained against a volunteer member and other career fire fighters who wish to remain anonymous but share similar concerns. He comes to you directly while off duty because the issue does not involve his company officer or battalion chief. He explains that the complaints began when the career member on the shift before his raised concerns about a volunteer fire fighter's skill deficiency with her company officer and later escalated the issue to her battalion chief when no action was taken. He emphasizes that various members have raised similar concerns numerous times with company and shift officers but that nothing has been done about it, and the ill-prepared volunteer members are still responding to calls.

1. If you do nothing, what will happen?

Inaction will allow the skills deficiency allegations to continue to spread. The sentiment can have a polarizing effect on the department, turning volunteer and career members against each other regardless of their training. A polarized department will compromise the present and future effectiveness and success of the entire organization.

Doing nothing would also implicate you as the officer responsible for the entire department operations if in fact your subordinate officers have received legitimate complaints and done nothing about them. Although the legitimacy of the complaints and whether any officers have been informed has not been verified at this stage, it is your responsibility to determine the facts of the situation. Your officers similarly have a duty to investigate complaints, take appropriate actions, and report results as necessary.

At the operations chief level, inaction also significantly increases the chance of the complaints going public. Regardless of the complaints' validity, the group of career members involved in the situation may be apt to share their claims with the public through print or television news outlets or social media. As an operations chief you have a duty to address operational issues while simultaneously limiting the department's negative exposure in the media. Even if the allegations are unfounded, once they have been aired in the media, the fallout will have to be managed. Doing nothing in this scenario could prove costly for the organization.

2. What do you say or do?

As in any situation, your first step is to gather the facts. First identify what, if any, subordinate-level officers received the complaint and what, if anything, they have done about it. When questions of skill competency arise within the ranks, there are potential safety concerns for the department's members and the public. As operations chief you must assess the claims' validity. Additionally, you must resist the temptation to step in and address these issues directly. There may be appropriate opportunities for direct involvement in the future, especially if the allegations go public, but for the time being you must focus on assessing the situation through your shift command officers and company-level officers. Make them aware that you know about the complaints and that you want to see a resolution through their actions. This is important because it maintains the focus of attention where it should be: at the level of the organization closest to the alleged problems. You may need to provide guidance or establish objectives, but you must manage the department's operations through your officers, not in place of them.

You must also investigate why subordinate levels of officers have allowed any claim of skill deficiency to reach the operations chief level. There should be a system of regular training and supervision in place to easily and soundly discount claims of ill-prepared fire fighters, regardless of career or volunteer status. A company officer or shift command officer who has heard such claims should constantly identify training opportunities for all personnel. These officers should take corrective coaching actions for any member, career or volunteer alike, who may show even the slightest sign of struggle in accomplishing their assigned functions. If there is any truth to the skill deficiency claims, there is an additional issue to address: your company and shift officers' supervisory skills. There may be actions that were taken or process improvements under way but as the operations chief you must determine what those actions have been up to this point.

You will also want to recognize and assess the safety factor in these allegations. There is a difference between fire fighters subjectively falling short of another member's expectations and if the fire fighters are truly unable to stretch a hose line into operation, properly connect to a hydrant, or avoid walking through a chemical spill. The competency issues may require the same actions, but the speed of action may be adjusted depending upon the specific details and severity of the allegations. Even so, remember these are allegations and at this point you want to be cautious about jumping into an issue that may turn out to have little or no validity.

3. To whom do you say it?

As operations chief you will need to quickly communicate up and down the chain of command. First inform the fire chief of the details you have received, the actions you are preparing to take, and the time frame in which you might have more information for her. You want to communicate with the fire chief first so she is informed and can subsequently inform any elected officials that may be pulled into a fire department issue. As an executive-level officer, you must consider issues your department leadership may face. The fire chief's supervisor, whether elected or appointed, works for the public and will feel compelled to have answers for the public if confronted with a public complaint about the issue. You want to provide him or her with the ability, if necessary, to confidently assure the public that he or she is aware of the situation and has confidence that the fire department is handling the matter.

Next, communicate with your direct report officers, as your shift command officers need to know what you have heard. You could conduct a staff meeting with all the shift command officers at once, but since multiple shifts are involved, you will likely need to have the same discussion multiple days or nights in a row when the individual shift commanders are on duty. It is advisable to hold face-to-face meetings rather than communicate through email or memo correspondence, because they allow you to fully assess the situation by observing facial expressions, eye contact with other officers, and so on. Indicate the questions you want your officers to answer and the time frame for reporting their findings back to you. You may ask your commanders when the last skill assessment was conducted for all shift personnel, what observations they have made on fire scenes regarding performance of duties by all personnel, whether any personal agendas may be influencing these allegations, whether the shift command officers have been notified of these allegations and, if so, what actions they have taken to mitigate the situation, and whether any company officer has made these allegations against volunteers. Even if the allegations prove false and the volunteers in question are capable and unjustifiably targeted, a consequence for false allegations will need to be administered.

If you meet with the shift commanders individually rather than as a group, it would be helpful to summarize the individual responses and share them with your shift commanders. As operations chief you want to take these results, positive or negative, and have the shift commanders engaged in reinforcing the positives and correcting the deficiencies. Sharing the summarized results also assists those shift command officers who may not have an issue now to be aware of potential issues so they do not develop.

4. What are two other alternatives you could use?

Consider a skills competition for all members. You may design individual or team events consisting of basic firefighting skills performance. Times may be kept for ranking purposes or pass/fail criteria may be established. Over the course of weeks or months, have members participate in these skills competitions and assess performance directly. The advantage to this alternative is that it keeps false allegations from being easily dismissed. It sends the message that a skills deficiency allegation is treated seriously but that everyone must be ready to perform, not just an identified subgroup within the organization. This alternative is not intended to punish the whistleblower but is intended to communicate that all personnel must be ready to perform required skills at all times.

A second alternative could be to assign your shift command officers, through their company officers, to sit down and talk with the volunteers identified in the allegation. This should only be considered after some initial fact finding is completed by the company or shift officers to see if the allegations have merit, partial or otherwise. If they do, have the company officers brief the affected volunteers and ask if there are training improvements or process improvements that could assist them in any way. This demonstrates to the volunteers that the organization is not quick to blindly act on allegations without substantiation and that the department wants to meet needs where they exist without blame or ridicule. If this option is selected, company and shift command officers will have to be cognizant of shift culture to ensure it does not become further polarized. You do not want the group of accused volunteer members seeking out the career members who made the allegations.

5. What, if any, legal issues should you consider?

The operations chief level of the organization has a duty to ensure the safe and effective operations for the entire department. Safety and performance allegations should be fully investigated. This investigation does not have to be conducted by the operations chief directly but must be completed to avoid charges of negligence. If a complaint was raised and dismissed without full investigation, the supervising officers run the risk of negligence in their duty to provide safe and effective operations within the department. Should that occur, allegations and degrees of correctness will likely be challenged in the courts, but the legal issue to think about is the operations chief's duty to determine fact versus fiction. This is especially important because officers supervised by the operations chief were allegedly made aware of the problem and did nothing about it.

6. What is your time frame for addressing this issue?

You will want to act swiftly but not emergently. When performance and safety allegations are involved you cannot afford to have weeks pass before taking action. You may be able to develop a list of questions and possible actions to take in a matter of a day or two and then communicate that list of questions over the following period of days. You will need to allow subordinate officers time to be briefed on the issue and then carry out their investigations in a timely manner within existing scheduled activities. This could be a shift or two or could be longer if other preexisting organizational commitments exist.

7. What is your last-resort action?

The essence of this scenario is a skill proficiency issue with an associated training issue. The responsibility of ensuring skills are presented, practiced, and polished falls to the company officer. To rectify a proficiency issue, assert your expectations to the shift commander and the company officer that basic skills are to be retaught to all personnel under their command and these skills are to be performed and evaluated for all personnel and the results documented and forwarded to the shift commander for summary presentation to you, the operations chief. Notify the shift command officers that skill deficiencies will be considered a direct reflection on their performance of duties to supervise the shift to ensure response readiness. Inform your officers that though the allegations may be unfounded, due to the fervor of the complaint and the significance of the allegation compromising the service to the community and the potential safety of fellow members, you must provide a definitive and defensible result to the questions raised. Instruct shift commanders to have these training reviews and assessment result documentation to you within three to four weeks. This allows them adequate time to train members and have them practice skills before assessing skill performance. Hold your shift commanders accountable for this action.

Note that your action does not have any direct contact with the career members, the volunteer members, or the company officers. Your last-resort action also does not involve taking a hands-on role to addressing the issue. At the operations chief level, you must leverage organizational results through other officers. Task them and let them carry out the request. You have an obligation to hold them accountable for assignment completion and for assessing the documented results they bring to you, be those proficient or poor.

8. If you reach your last-resort action, what effect will you be dealing with two weeks after taking action?

There may be disbelief that you have not engaged directly. Those complaining will want to see you validate their complaints by taking action against the skills-deficient volunteer members. The career members may be frustrated or angry and feel that you do not care enough about the issue to intervene. You will probably have many members complaining about the fact that the complaint resulted in requiring all fire fighters to complete review training and participate in skills assessments. Size up the station issue here: Fire service members often feel that they proved their skills and abilities when they went through the recruit academy and that these skills are refined and polished by experience on the street every day. Even if a department makes only a handful of calls in a month, many fire fighters may believe that this deems them too busy to need to demonstrate to anyone that they can still perform such skills. They may point to the fact that there are no ongoing conflagrations around the community as proof enough that they have always performed necessary skills correctly and efficiently. Some members may feel that asking them to participate in a skills assessment is unacceptable enough; mandating it because of some complaints will likely result in intense anger and criticism from veteran members. Be proactive and consider having shift commanders encourage members to view the skills assessment as an opportunity to show off what they can do rather than an exercise in re-proving their skills and abilities.

2. Fire Fighter Fraternizes with City Manager

▸ NFPA 1021: 5.2.1, 5.4.6, 6.2.1, 6.2.7, 6.4.6

Your battalion chief stops by your office and asks for a moment to brief you on a situation that his station officer shared with him last shift: During routine physical fitness training, a fire fighter and the newly elected labor representative, Kevin, mentioned to the station officer that he has become increasingly friendly with the city manager. The lieutenant did not specifically know what to do or say so he mentioned it to the battalion chief to keep him informed. Your battalion chief tells you that he, too, is unsure of what to think of the situation but thought you should be aware of it.

Your battalion chief explains that Kevin did not previously know the city manager but that since Kevin's election, the city manager has extended dinner invitations to him and given him tickets to sporting events. In some instances, the city manager and Kevin have been the only city personnel in attendance. The city manager's family and Kevin's family have also attended multiple family picnics together. From what Kevin shared with his station officer, your battalion chief knows that fire department business sometimes enters Kevin's off-duty conversations with the city manager.

1. If you do nothing, what will happen?

Although there may be very little you can do in this situation, if you do nothing, you will be subject to whatever may unfold should the relationship affect the fire department's operations. The city manager could potentially begin using Kevin as an informal source for information about the fire department. Concerns from the line fire fighters, failed policy implementation, and information about the fire department administration are all potential topics that could easily come up in casual conversation but ultimately jeopardize the integrity of the fire department's chain of command.

Alternatively Kevin could intentionally broach fire department issues in the hope of influencing how the city manager directs the fire chief or fire administration. You must act within your scope to prevent this type of organizational issue from developing.

2. What do you say or do?

If your department already has a policy limiting personnel interaction with political leaders you may have a point of reference, but it is more likely that no such guidance exists. This scenario is particularly tricky because it involves circumstances outside of the workplace, which are generally off limits to department policy. Additionally, the possible motives behind the friendly activity makes the scenario challenging. Having a close friendship with someone in a position of influence can be attractive. Since Kevin is the newly elected labor representative and the city manager is an elected official, the friendship could be a strategic move for either party to gather information. The motive could be pure; simply two individuals with similar interests developing a close friendship. However, it cannot be overlooked that the two parties' occupational positions could create an opportunity to exert influence on the other at work, such as pushing a particular agenda for the fire department's future or an effort to encourage or undermine existing leadership. The crux of the issue is that you do not know why the social interactions began and you are not likely to find out. Be a part of the process to keep communications open and kind. In this type of situation, there is likely little else you can do.

3. To whom do you say it?

Communicate with your supervising officer, who in most organizations is likely to be the fire chief. Let the chief know what you do and do not know. Keep communications honest and unbiased. Remind the chief that the scenario could be nothing more than two individuals becoming close friends who just happen to have day jobs that are tied to the fire department. Take care not to jump to the conclusion that some political undermining is taking place—such assumptions may harm your department. When you are at the executive level, leading an entire fire department operation, you must develop the self-discipline to avoid being influenced by rumors, perceptions, and speculation. Although these things are likely to come about over time from the general membership, they cannot come from you as an executive-level officer.

In the event you are asked about the situation from the general membership, respond within the chain of command. Provided there are no specific policies in place forbidding such activity, the message of response should be: "The department is aware of the situation, and personal situations outside the department are of no concern within the department's operations. In the event a personal situation develops into an operational matter, the appropriate personnel within the department will address the situation." Such a message will likely provoke responses such as, "What will be done? How will it be addressed? Who will deal with it?" Demonstrate restraint; these follow-up questions do not need to be answered because they concern a personal matter. Instead, an appropriate response—that demonstrates the organization respects the privacy of its members—is: "The appropriate people are monitoring the situation."

4. What are two other alternatives you could use?

One alternative might be to have a brief, private conversation with Kevin, the member of the fire department involved. Explain that you have been made aware of the social relationship with the city manager and her family and that the relationship opens the door to speculation about the affect such a relationship could have upon the fire department. Reassure Kevin that the relationship is not open for comment, approval, or disapproval by the department or its membership. However, inform Kevin that the fire department membership has noticed these interactions and that people within the department are curious and may not be willing to accept an honest friendly interaction at face value. You may want to specify that Kevin has nothing to be concerned about from your perspective or the department administration.

A note of caution: If you previously communicated with your supervisor and he or she suspects an ulterior motive, do not have this conversation with Kevin. If there is concern of an ulterior motive, regardless of who is believed to possess it, then quietly monitoring the situation may be the best course of action. Should Kevin ask you whether anyone in the fire department has taken issue with the situation, you would face the quandary of telling the truth and creating anxiety with Kevin while exposing your chief's concerns or remaining silent, which is untruthful and tarnishes your integrity as an officer and a person. Avoid the dilemma and do not have the conversation with Kevin if your superior officer suspects ulterior motives for the newfound friendship.

A second alternative might be to monitor the situation and keep account of fire department issues that may receive increased attention from the city manager. This could be positive or negative attention, but by attending to potential effects, you can assess possible causes. Again, even if there are correlations that could be traced back to these increased social interactions there will be limitations to what can be said or done in the workplace. Tracking these increased interactions with the city manager or city officials in general, however, provides data for any conversation in the future. Even if the conversations revolve around perceptions more than any real actionable consequences, the data will be helpful to keep the conversation objective.

5. What, if any, legal issues should you consider?

The biggest legal issue you will have to consider is civil rights, specifically freedom of speech and freedom of association. Though the workplace creates channels for supervision and occupational direction, the workplace is not entitled to direct personal activities and relationships. Even where departments have policies in place prohibiting spouses from working for the same department or division, the choice of marrying an existing co-worker still rests with the individuals involved; the policy simply states that the consequence is for one of the two employees to resign. This example serves to remind you that you cannot simply approach your member and direct him to sever the friendship.

6. What is your time frame for addressing this issue?

This issue is not an urgent one. In fact, considering your alternatives as time unfolds will actually help clarify any issues. You do not want to leave this issue uncommunicated with your supervisor for very long after you have become aware of it. However, any action alternatives discussed can be considered over weeks or even a few months.

7. What is your last-resort action?

In this scenario a last-resort action may be to brief your fire chief and allow her to address the subject with the city manager if she so chooses. Because of the sensitive nature, the limited consequences on the daily operations of the fire department, and the recognition of your limited ability to influence the situation one way or another, you may determine this issue needs to be forwarded up the chain of command to the next level for consideration. This is a last-resort action because, for the many reasons already discussed, you must recognize when a particular situation is bigger than your level of responsibility and capability to address it. Using the emergency scene size-up analogy, there are times when you are faced with enough information to recognize that an event is bigger than your response. Maybe despite seeing cars in the driveway, the dense black smoke pushing against all the windows of a one-story house upon your approach causes you to go defensive rather than enter and search. Such a size-up decision is based on the knowledge that any occupants inside have already perished, despite your not knowing for sure whether anyone is inside. Similarly, this personnel situation may require you to go defensive rather than take an active offensive approach. Consulting with your chief is prudent in this case, especially if the previously discussed alternatives have been attempted with no results.

8. If you reach your last-resort action, what effect will you be dealing with two weeks after taking action?

The reaction to this particular last-resort action may be very little if any at all. Because the issue is based on relationships and activities outside of the workplace, there is not likely to be significant rumors or fallout from the alternative. The fire chief may decide to wait and see what happens over time. Even if the fire chief decides to discuss the matter with the city manager, it is not likely to have an adverse impact on the operations of the department. It may have some strained relations with the fire chief, depending on the culture of the city or the personality of the city manager, but there are too many variables to identify specific consequences here.

3. Citizen Complaint: Transport Delay

▸ NFPA 1021: 5.2.1, 5.4.6, 6.2.1, 6.2.7, 6.4.6

You receive an email from a citizen alleging that one of your ambulances, Ambulance 7, experienced navigation issues en route to the hospital, which resulted in delayed patient treatment. The email is from the patient's family and states that the ambulance driver did not know where he was going and that, as a result of the long travel time to the hospital, the patient suffered adverse effects, including delayed surgical intervention and spread of infection.

You have asked for a report from the dispatch center regarding the Ambulance 7 incident times for that particular call. A quick review of the computer-aided dispatch times revealed there was a 10-minute longer transport time than normal for the destination hospital.

1. If you do nothing, what will happen?

Failing to act in this situation would put the department in jeopardy. Whether the complaint is accurate or even valid will come to light soon enough, but you have no way of knowing whether this is the only notice your department will receive. You must act to ensure appropriate investigation can be completed and legal preparations can be made should the citizen file a legal action against the department.

2. What do you say or do?

Recognize this is a department operational issue and is ultimately your responsibility as the operations chief. The people working in the various divisions, shifts, stations, and companies are extensions of you. You should treat the complaint as a breach of departmental effectiveness until proven otherwise—take care not to interpret this statement as guilty before proven innocent. There may be good reasons behind the transport delay, there may be misunderstandings by the citizen, and there may not even be any truth to the complaint; however, you must treat it as a serious deficiency in the department's operation to provide each customer with a satisfying interaction.

Your first step should be to communicate. Completely inform your supervisor with the information you received, including dates, times, sources, and any initial steps you intend to take.

Next, investigate. Determine the facts of this incident, such as the names of the personnel on the Ambulance 7 response crew; the response times; the stations and units involved; any pertinent road construction issues, traffic delays, or vehicle maintenance issues; and the training of responders involved. These are just a few of the many elements to investigate and gather facts about.

An important aspect to consider at the operations chief level of the organization is that, regardless of the complaint, you need this information to assess the effectiveness of the department's daily operations. Do not overreact and create a volatile situation. Just as when you size up a fire, you do not run around chaotically, yelling, "Get the hose! Get the ladders! Save the children!" Similarly, do not size up this issue by bringing the Ambulance 7 crew into your office and exclaiming, "What in the world happened on this run?"

Work through your subordinate officers initially. If you receive feedback that merits your involvement be ready to interact but allow your shift and company officers to do their work. If this complaint is valid because someone simply did not know where he or she was going, it is an issue for the individual driving Ambulance 7, Ambulance 7's company officer, and the company officer's shift supervisor. Training, supervision, and guidance are all areas of responsibility that reside closer to the response level than the operations chief level, so identify the questions that need to be answered and send them down the chain of command. You will need tò stipulate the deadlines for the officers to have their answers to you.

Once these actions are under way, assess from your supervisor what, if any, additional actions you need to take. Next, prepare by anticipating potential responses so you are able to process subsequent actions based on the results your officers give you. There are a few basic outcomes to anticipate.

The simplest outcome is that the citizen is inaccurate about the response to the hospital. If facts reveal that this particular Ambulance 7 run was within the usual response time parameters, and the times are accurate, normal, and operationally acceptable, then you have a public relations issue to contend with. If the complaint is valid and the Ambulance 7 response was unusually long or delayed, you may have a coordination issue with outside agencies such as the hospital or a public works department. You may also have an internal performance issue or possibly some combination of both issues to address. If you have a delayed Ambulance 7 response with the causes for such delays known prior to the run, you have both a performance and a supervisory issue to address. Though the facts are not yet known, you can consider how you would deal with each of these possible outcomes so that you are be prepared to take action when the time comes.

3. To whom do you say it?

Discuss the complaint with your supervisor, who is likely to be the fire chief. Next, seek the most qualified person to gather answers to questions. For a response data question, you might ask the information technology or incident report quality review personnel. For an account of the run from the crew involved, you will need the shift supervisor on duty that day to ask his Ambulance 7 company officer to ask the Ambulance 7 crew involved about that particular run. To investigate potential traffic or construction issues, make sure you have information regarding the exact route taken by Ambulance 7 so you can ask the public works or street department manager about construction in the areas identified. As operations chief you will likely supervise the training division or training programs. Ask the training officer about driver training programs and recordkeeping generally. Ask the fleet maintenance department or the maintenance person in the fire department about any maintenance issues specific to the apparatus involved in this response so you can determine if mechanical issues may have played a role.

4. What are two other alternatives you could use?

One alternative might include asking the communications center to send you a copy of all the response records for this particular incident, including computer-aided dispatch logs and radio communications recordings, if possible. Listen for yourself how the incident progressed. Did radio messages indicate concern, indifference, or urgency? Note that these types of sources are likely to be requested if a legal action is filed so it is advisable to listen for yourself early so that you may assess how such data could affect the department. Remember you want to address the complaint but you also want to ensure appropriate service delivery to your customers at all times and under any circumstances.

A second alternative should be carefully considered with your fire chief and the legal counsel for the department before acting. You may, with permission of the fire chief, set up a face-to-face meeting with the person who sent the email complaint. This alternative can be risky because although the meeting could be perceived by the complainant as responsive to his or her concern, it could also be perceived as a show of force to intimidate the complainant. A second risk is that comments made during the meeting can be misunderstood or even used against the department should legal proceedings be brought against it. This meeting should be attended by you, the fire chief, the department's legal counsel, and possibly the shift supervisor for the Ambulance 7 crew from that day present with the person sending the complaint. Do not bring the individual Ambulance 7 responders involved into the meeting because you do not want to create a scenario of potentially heated direct questioning.

In the meeting, acknowledge that the complaint was received and that the department is taking the matter seriously and investigating it fully. This does not necessarily mean legal action will be avoided or that you will hear nothing further from the complainant after the meeting. However, there are times when acknowledging a complaint and reassuring the complainant that you will take steps to prevent future occurrences are all that the person really seeks. You may be able to close the issue right then, at least from the complainant's point of view. However, you still have a responsibility to the department to ensure that if in fact a mistake occurred, a remedy is enacted to prevent future occurrences.

5. What, if any, legal issues should you consider?

There are many issues that could potentially be brought against the department. Issues could include negligence, improper training or maintenance, failure of supervisory personnel to ensure adequate performance of duties, or the patient's family seeking financial damages for losses to the patient or patient's family. Note that in this scenario, the documented complaint would be available for public inspection through a Freedom of Information Act request and could be used by the media, other legal actions past, present, or future, as well as trial by the court of public opinion. Recognize how severe this issue can become and how it can impact the department. After informing your supervisor, seek legal advice early.

6. What is your time frame for addressing this issue?

Do not delay in addressing this matter. After receiving the complaint, take a few minutes to make sure this incident was in fact your department's response. After confirming the incident response in question was your department's call for service, it is recommended you communicate with your supervisor immediately. A quick phone call instead of an email or memo keeps your fire chief briefly informed. The fire chief may have instructions for you or may want you to start the investigation with regular follow-up communications.

When you communicate to your shift supervisors and company officers, you will want to identify deadlines for their fact finding that fall ideally within one or possibly two shifts of the involved Ambulance 7 personnel working at the fire station. This is not an issue you want to span over several weeks, because if legal action is taken against the department or its Ambulance 7 members, you want to demonstrate your willingness to tackle the issue as soon as you possibly had knowledge of any concern.

7. What is your last-resort action?

Your last-resort action may be to bring in the Ambulance 7 crew involved for direct questioning with you and possibly legal counsel. This is a last resort because dealing directly with the operations chief may put the Ambulance 7 members on the defensive. Second, it circumvents the effectiveness of the chain of command, which can disenfranchise your subordinate officers. Third, by intervening directly, it takes your attention away from the functions of managing the department's operations. In addition to your regular meetings, appointments, and functional responsibilities, you will be largely responsible for conducting the investigation. It is encouraged that you delegate as much of the investigation as is practical and satisfactory to your fire chief and legal counsel.

8. If you reach your last-resort action, what effect will you be dealing with two weeks after taking action?

You will have an increased workload as you coordinate and conduct elements of the investigation. Your direct involvement with the Ambulance 7 response incident will likely result in increased attention from the membership, as the operations chief is normally not engaged with individual responses. This could create speculation, rumors, and possibly even animosity. Such animosity may be directed toward the fire department administration for not believing in the street responders, against the public for feeling that the fire department's services are insufficient, or possibly a bit of both. You and your officers will not be able to discuss much about the situation or dispel the rumors and inaccuracies. Because this particular scenario could have pending legal action, discussing the specifics of the investigation or the Ambulance 7 response may be inappropriate or compromise the legal case.

CHAPTER 7

Fire Chief Scenarios

1. Citizen Complaint: Transport Delay

▸ NFPA 1021: 5.2.1, 5.4.6, 6.2.1, 6.2.7, 6.4.6

You receive an email from a citizen alleging that one of your ambulances, Ambulance 7, experienced navigation issues en route to the hospital, which resulted in delayed patient treatment. The email is from the patient's family. It states that the ambulance driver did not know where he was going and that, as a result of the long travel time to the hospital, the patient suffered adverse effects, including delayed surgical intervention and spread of infection.

You have asked the operations chief for a report regarding the Ambulance 7 incident times for that particular call. A quick review of the computer-aided dispatch times revealed there was a 10-minute longer transport time than normal for the destination hospital.

1. **If you do nothing, what will happen?**
2. **What do you say or do?**
3. **To whom do you say it?**
4. **What are two other alternatives you could use?**
5. **What, if any, legal issues should you consider?**
6. **What is your time frame for addressing this issue?**
7. **What is your last-resort action?**
8. **If you reach your last-resort action, what effect will you be dealing with two weeks after taking action?**

FIGURE 7-1 A citizen complaint may be received by the fire chief directly, from a department website, or through another executive-level officer within the fire department. When the complaint is addressed to you, you will own responsibility for handling it.

Courtesy of Jeff Barlow.

2. Fire Fighters Critical of City Administration

▸ NFPA 1021: 5.2.1, 5.4.6, 6.2.1, 6.2.7, 6.4.6

A vocal minority of fire fighters in your department have been complaining to the press and making public statements about the city administration's poor treatment of fire department personnel. The complaints started many months ago in a fire station, then spread to more stations and shifts, and over the past 3 months—although the responsible group is limited to a minority of the entire membership—they have increased in intensity. The group accuses the city administration of not valuing or respecting the fire fighters' work, citing modest pay raises of less than 2 percent, no hiring of additional fire fighters to increase staffing levels, and deteriorating fire stations, equipment, and apparatus. The complaints have become increasingly public, with opinion pieces in the local newspaper and campaigns on Facebook and other social media sites dedicated to exposing the fire department conditions and the alleged lack of city hall support. The increased publicity seems to be designed to turn public sentiment against the elected officials and put pressure on you and your municipal colleagues to increase fire fighter pay, make renovations, and replace old apparatus.

During a regularly scheduled department-head meeting with the mayor and the city manager, the city manager states she has had enough of the constant complaints and accusations from the fire fighters and tells you and the mayor it is time to put a foot down and show the fire fighters who they really work for. She suggests that it might be time to consider fire department staffing alternatives that would provide the same service with fewer fire fighters per unit or rethink the priorities of the city's public safety needs. You know that the city manager's comments are borne out of frustration about the fire fighters' continued complaints and accusations. Although you encourage the mayor and city manager to not overreact, you are keenly aware the position of fire chief serves under the sole authority of the mayor. You are also aware that in the past 24 months, there have been slight salary increases, the replacement of one fire engine and one ladder truck out of the 25-apparatus fleet, and renovations to the 75-year-old firehouse. In addition, you know that the city is facing significantly reduced revenues with no significant improvements within sight.

1. **If you do nothing, what will happen?**
2. **What do you say or do?**
3. **To whom do you say it?**
4. **What are two other alternatives you could use?**
5. **What, if any, legal issues should you consider?**
6. **What is your time frame for addressing this issue?**
7. **What is your last-resort action?**
8. **If you reach your last-resort action, what effect will you be dealing with two weeks after taking action?**
9. **What organizational impacts might the last-resort action have on the fire department?**
10. **What organizational impacts might the last-resort action have on the city administration?**

CHAPTER 8

Fire Chief Discussion

1. Citizen Complaint: Transport Delay

▸ NFPA 1021: 5.2.1, 5.4.6, 6.2.1, 6.2.7, 6.4.6

You receive an email from a citizen alleging that one of your ambulances, Ambulance 7, experienced navigation issues en route to the hospital, which resulted in delayed patient treatment. The email is from the patient's family. It states that the ambulance driver did not know where he was going and that, as a result of the long travel time to the hospital, the patient suffered adverse effects, including delayed surgical intervention and spread of infection.

You have asked the operations chief for a report regarding the Ambulance 7 incident times for that particular call. A quick review of the computer-aided dispatch times revealed there was a 10-minute longer transport time than normal for the destination hospital.

1. If you do nothing, what will happen?

If you do nothing you risk condoning an attitude of indifference across the department. As the fire chief of an agency that exists to serve the public, you must recognize the need to respond and react to the complaint. This particular scenario has a very real potential to become a legal matter, and if you do nothing, you risk making a legal case worse for the department. If you do not take action, the complainant may escalate the issue within the organization. You do not want a member of the city's administration, the board of directors, or the larger parent organization learning that a complaint regarding emergency services was expressed to the fire chief and never addressed. Equally detrimental would be if they learned of the complaint from a citizen rather than you, the executive officer who had knowledge of it.

2. What do you say or do?

First, acknowledge receipt of the complaint. Do not attempt to explain, excuse, or justify the situation. Taking the time to confirm the message was received can help calm the situation by signaling that the complaint will not simply be disregarded or covered up. Next communicate with your chain of command. Your supervisors—be it a city manager, the mayor, or a board of directors—will need to be informed of the complaint and what steps are being taken. This is important because if the fire chief has not provided the complainant with satisfactory results, he may choose to escalate the issue. Since elected or politically appointed officials, such as your supervisors, often interact with community members in public meetings, which may be televised, avoid any risk of your supervisors first hearing about the situation under these circumstances.

After communications have been made, it is important to investigate the facts of the situation. Make no assumptions. As fire chief, you have a responsibility to the citizens you serve to ensure effective service; however, you also have a responsibility to support your department's members and stand up for them when they are wrongly accused or have given their best possible effort despite a negative outcome. Neither stakeholder can be dismissed. Investigate the facts to determine what next steps need to occur. You may delegate the investigation to an officer closer to the information, such as an assistant chief, district chief, or battalion chief. However, be careful not to put subordinate officers in a position where they are investigating their own crews or shifts. If the Ambulance 7 crew works out of Station 7, you may not want the Station 7 company officer leading the investigation. The company officer may be involved with the investigation but is likely not the best choice to lead it.

3. To whom do you say it?

As previously mentioned, communicate with the complainant, your supervisors, and the person identified to lead the investigation, unless you decide to investigate the matter personally. You will also want to communicate early with the department's legal counsel. This step may be at the direction of your city hall contact, but regardless, you will want to share the information you receive with legal counsel so they can advise you appropriately on what actions to take.

4. What are two other alternatives you could use?

One alternative could be to directly contact the Ambulance 7 crew involved in the incident. Request that they stop by your office the next duty day they are available, and ask them to recall the details of the response. This is one of the most direct investigative routes, but it will be difficult to downplay any resulting hype or stigma about reporting to the fire chief to explain their actions on a run. It could provide insight to any obvious explanations, such as road construction, mechanical issues, or destination change mid-transport to the hospital.

A second alternative might be to have the operations chief, EMS chief, or battalion chief investigate other EMS transports from other crews, stations, and shifts to see whether the transport time for this incident is atypical or aligns with the response norm, contrary to what the complainant may perceive. It is important to recognize that the investigation and results may uncover a system-wide issue otherwise unnoticed until the complaint was received. A system-wide issue may not necessarily be due to an underperforming crew on Ambulance 7 or other crews but may be an issue in need of further study and problem solving to prevent future complaints from the public served. Global Positioning System equipment or mapping software for mobile computers may need to be considered to assist systematically in reducing transport times or even response times. Although these items may be expensive, they may be a cost-effective alternative to litigation or a judgment against the department. Determine the cost versus benefit.

5. What, if any, legal issues should you consider?

If the complainant alleges that a delay in treatment at the facility resulted in a poor outcome for the patient, there is a strong potential for litigation. Every action you do or do not take will be scrutinized. Every communication you make—especially email, memo, text, or letter—can be requested under a Freedom of Information Act request. If this issue becomes a lawsuit, you will have to explain and defend every action or inaction and every conversation within the investigation. Be sure to involve legal counsel at the beginning. Response times, computer-aided dispatch tapes and logs, and expert testimony from outside your department may be issues to contend with so be sure to investigate thoroughly before responding to the complainant. The department, the city or board, and the legal counsel cannot afford to be surprised by data discovered in court should the complaint become a lawsuit.

6. What is your time frame for addressing this issue?

Communications should occur within the day you receive the complaint. The investigation itself can be a matter of urgency but not emergency. The initial steps of the investigation should take place within the week. Take follow-up measures as soon as practical. Draft communication back to the complainant within a day or two of receipt of the complaint, and have legal counsel review it before it is sent. This prompt reply assures the customer that his complaint has been received and considered. However, as legal counsel will likely inform you, it potentially starts the clock for a timely resolution or response should the complainant file a legal action.

7. What is your last-resort action?

The last-resort action would be a letter of apology if the investigation revealed poor performance on an isolated response by Ambulance 7. This becomes extremely sensitive and will require expert advice from your legal counsel. If the investigation is complete and errors were made, you must acknowledge that the complaint filed brought the matter to the department's attention, that corrections have been made to prevent future occurrences, and that you appreciate the person taking the time to inform the department of his experience so that continuous improvements to customer service can be implemented. It is not appropriate to discuss specific personnel corrective actions and may not even be appropriate to discuss the specific systemic corrections being implemented. If this last-resort action is reached, be sure your elected or appointed supervisors and the legal counsel sign off on any correspondence before it is sent.

8. If you reach your last-resort action, what effect will you be dealing with two weeks after taking action?

Because the public consists of many different perspectives, it is difficult to anticipate exactly how the complainant will respond. Some may be pleased to have a letter of apology. Some may even be impressed that the department acted upon the complaint and was willing to admit there was a problem that was then addressed. Others will use this letter as a reason to take it further either in the courts or to draw more attention to the issue publicly. They may see the letter of apology as an indicator of financial damages to be easily gained.

As a fire chief you may not know exactly how this last-resort action will impact the city administration's elected officials. However, be mindful that should this letter become a part of any legal proceedings that award financial judgment against the city or the fire department, budget cuts may be required to offset the fines or fees.

Your department may see a letter of apology as a betrayal of the personnel, specifically Ambulance 7. It could run the risk of eroding the trust or respect of the department battalion- or company-level leadership; however, it is not likely that the department membership will have any details of the correspondence unless that citizen goes public. Remember that once the letter is sent, how it is received is out of the department's control.

2. Fire Fighters Critical of City Administration

▸ NFPA 1021: 5.2.1, 5.4.6, 6.2.1, 6.2.7, 6.4.6

A vocal minority of fire fighters in your department have been complaining to the press and making public statements about the city administration's poor treatment of fire department personnel. The complaints started many months ago in a fire station, then spread to more stations and shifts, and over the past 3 months—although the responsible group is limited to a minority of the entire membership—they have increased in intensity. The group accuses the city administration of not valuing or respecting the fire fighters' work, citing modest pay raises of less than 2 percent, no hiring of additional fire fighters to increase staffing levels, and deteriorating fire stations, equipment, and apparatus. The complaints have become increasingly public, with opinion pieces in the local newspaper and campaigns on Facebook and other social media sites dedicated to exposing the fire department conditions and the alleged lack of city hall support. The increased publicity seems to be designed to turn public sentiment against the elected officials and put pressure on you and your municipal colleagues to increase fire fighter pay, make renovations, and replace old apparatus.

During a regularly scheduled department-head meeting with the mayor and the city manager, the city manager states she has had enough of the constant complaints and accusations from the fire fighters and tells you and the mayor it is time to put a foot down and show the fire fighters who they really work for. She suggests that it might be time to consider fire department staffing alternatives that would provide the same service with fewer fire fighters per unit or rethink the priorities of the city's public safety needs. You know that the city manager's comments are borne out of frustration about the fire fighters' continued complaints and accusations. Although you encourage the mayor and city manager to not overreact, you are keenly aware the position of fire chief serves under the sole authority of the mayor. You are also aware that in the past 24 months, there have been slight salary increases, the replacement of one fire engine and one ladder truck out of the 25-apparatus fleet, and renovations to the 75-year-old firehouse. In addition, you know that the city is facing significantly reduced revenues with no significant improvements within sight.

1. If you do nothing, what will happen?

Inaction will allow the level of complaining to increase and risk further alienating the fire department, which could result in unpleasant operational changes should the city manager influence the mayor to reduce fire department staffing. Doing nothing also leaves the city and the fire department open to an increased volume of public attacks. It could be used to support the claim that the fire chief is aware of the situation and is not fighting hard enough with the city administration to bring about needed changes.

2. What do you say or do?

Encourage the mayor and the city manager to consider that the fire fighters and the city administration work for the people of the community. Focus on some of the positive aspects of the department and the services it provides. Acknowledge the positive actions the city has taken to improve fire department service levels in the community, such as the new fire apparatus and station renovations. Next, clarify the complaints and accusations the city manager is referring to and the source of the information. It can be helpful for you to hear her perspective and sources. This may reveal the extent to which the news headlines are influencing her reaction. If you are able to learn of her sources, you may be able to more effectively intervene with the negative influences within the fire department. Underscore that the vocal group is a minority fraction of the department and that it would be unwise to create department-wide impacts to city service due to grievances from a minor portion of its stakeholders.

3. To whom do you say it?

These discussions should occur privately between the chief elected official, the city manager, and you, the fire chief. You are responsible for communicating to and with the fire fighters. If any outside inquiries come to you, you are responsible for remaining consistent in the message that the fire department is ready to serve the public when called upon.

Next, consider meeting with some or all of the identified fire personnel who are contributing to the situation. Inform those members that their actions are setting the stage for unpleasant and possibly unintended negative consequences for both the department and the community. Though you risk emboldening some of these personnel, you can express executive orders for fire department personnel to bring any concerns, complaints, or issues to the fire chief through the department chain of command. Convey that taking issues public through the media or other means without following the chain of command will be considered a violation of department orders. This step is important because it does not prohibit fire personnel from expressing their displeasure or concerns but does help to prevent the damaging consequences of negative press, which may or may not be justified. This action will create additional workload for the fire department chain of command; however, it will be effort well spent if it keeps fire fighter concerns and complaints within the department and prevents the related negativity from reaching the mayor or city manager outside of the chain of command.

4. What are two other alternatives you could use?

One alternative could be to assign your operations chief or battalion chiefs to investigate the source of the information reaching the city manager. Have these officers report their findings back to you so you can further evaluate the situation and assess whether further discussions with the mayor or city manager are appropriate. This alternative may run the risk of inflaming the members of the department who are uninvolved; it can create more negativity and possibly even turn members against one another. Consider also whether any chief-level officers are behind the public complaints or accusations. Upper-level officers should be aware that they are agents of a greater municipal organization and that any disagreement they have with the fire department or city administration should be broached in the proper place and time within the chain of command.

A second alternative could be to request that the city manager and mayor allow you to develop an informational media campaign to draw attention to the recent positive changes in the fire department. This campaign could highlight the apparatus replacements and station renovations and encourage future oversight to ensure the community is being protected effectively and efficiently. This alternative must be rooted in facts and not in embellished political promises or intentions. The media campaign would send a message to city hall personnel that the fire department supports a positive image for municipal services.

You may be cautious of a media campaign with a politically minded intent. In the fire service we spend much of our careers, volunteer or paid, facing tactical challenges with definitive outcomes: the fire goes out either because of our actions or because it burns up all the available fuel; either a patient lives or dies. The scenarios you face as a fire chief, however, are not so cut and dry. Fire chiefs and chief officers must operate an effective response operation in a political, rarely cut and dry, world of perceptions and people issues. These issues include political electability, special interests such as neighborhood associations and tax watchdog groups, and potential election challengers, which all influence the decisions mayors, boards, city managers, and other elected or appointed officials make. When considering a positive media campaign to offset any negative publicity the department may be facing, consider the mission of the department in context.

Regardless, do not compromise your personal integrity or the department's credibility. If you have nothing positive to work with, this alternative is not effective. Do not lie, mislead, or embellish just to please the city administration or counter negative commentary.

5. What, if any, legal issues should you consider?

There is always the concern that accusations of slander could be raised if fire personnel begin speaking out against individual city officials, such as claiming that a council person, city manager, or the mayor is threatening the safety of the citizens and trying to inflict harm on fire fighters by understaffing the department and withholding needed equipment. An unpopular decision made by an elected or appointed official is not grounds for personal attacks or fictional accusations created to try to further a cause. The law also has whistleblower protections, should a true unjust situation be exposed. It is difficult to know exactly what legal issues might be involved without further specific facts, but possibilities could include contract violations from labor or city management, or city or fire department policy infractions, if there are regulations forbidding public comment on city or department business without the proper use of the chain of command. Take care to consider legal challenges from the perspectives of the city and fire department as well as the vocal group of employees.

6. What is your time frame for addressing this issue?

If you are able to determine the source of the city manager's information, you can act swiftly. If a source is not readily identifiable, the time frame may become prolonged. Your ability to act may be limited because decisions made by the mayor or city manager are outside of your control. You may be able to influence them to lessen the impact on the fire department, but it is important to recognize once these policy makers decide to act, their actions and resultant consequence are not able to be undone. This size-up issue is comparable to an emergency scene size-up where an aggressive interior attack is believed to be best despite heavy fire involvement, only to have structural conditions become untenable and responders become severely injured. Although any incident commander would want to go back to the moments leading up to the decision and change the initial actions based on more complete size-up knowledge, it is not possible. In this scenario, fire personnel may fail to recognize that complaints and accusations can create unintended negative consequences for the entire department, such as budget cuts and reduced staffing.

7. What is your last-resort action?

In this scenario, your last-resort action may be to anticipate measures the city manager might implement as a result of the fire fighters' continued public complaints and to prepare specific reasons for why the city manager should reconsider. This is a last resort because it involves waiting for the city manager to react. For example, if the city manager tells you to reduce operational staffing from four to three members on an apparatus, you could counter the directive with various scientific time-work studies that have compared the effectiveness of fire operations for both three- and four-person crews.

Present your case for reconsideration as soon possible after receiving instructions, so it does not become an issue of insubordination but rather a team approach to preventing further damage to the public. This case should be prepared with indisputable facts and presented by the fire chief only. This step is important, because should the city manager choose not to reconsider, you have an obligation to work within your chain of command and roll out the department changes as if they were your own. Note that if the city manager gives a directive and advises it is not open for discussion, the directive should be followed.

Ideally you would want to exert influence in the flexible decision-making stage, but if the attempt proves unsuccessful, the last step is to respond to the reaction. For example, if the city manager decides there will be no pay raises for the fire department during the upcoming year, you will need to be ready to refocus your personnel to remain committed to the task and the department despite the lack of a pay increase for the year. This will require empathy and compassion from fire department leadership for the affected fire fighters, while continuing to communicate the mission of the department. Similarly, if the city manager decides there will be no further fire apparatus replacements until the public complaints cease and the financial concerns subside, you will need to make operational decisions about how best to reallocate your available resources. Such decisions are best thought through proactively rather than reactively.

8. If you reach your last-resort action, what effect will you be dealing with two weeks after taking action?

There is likely to be organizational turmoil. Members of the department who were not involved in criticizing the city's administration will be affected by the city manager's initiatives. Sides will quickly develop within the department as the group or individuals responsible for angering city officials are sought out. There will be significant hostility directed toward the fire department leadership for not preventing the city manager's reactionary measures. Dissenting personnel may become increasingly vocal, further exacerbating the issue between the fire department and with the city administration. This situation has the potential to leave a fire chief in a very difficult position: watching both the fire department membership and the city manager plan to retaliate against one another, all the while attempting to temper both perspectives and lead an organization.

9. What organizational impacts might the last-resort action have on the fire department?

The fire department will directly experience the consequences. Any decisions implemented as reactions to the initial issue will be systemic and potentially upset significant numbers of members who were previously uninvolved and possibly even unaware. The negativity may prompt these fire fighters to find fault with situations they might have previously otherwise dismissed—even though unrelated to the city administration's support of the fire department. As the turmoil builds, it can jeopardize the department's ability to deliver its services.

10. What organizational impacts might the last-resort action have on the city administration?

The city will soon realize the negative impact across far larger numbers within the fire department. As more personnel are upset, they are likely to seek public outlets for their issues. Rife with controversy, the city will have to dedicate more staff time to answering questions and explaining the situation and expend effort to control the negativity that may arise from vocal fire department supporters in the community. The city administration may communicate with increasingly strong language and actions that it, and not the complaining fire personnel, ultimately decides what is best for the public. This could result in increased time frames for pay freezes, continued sparse apparatus replacements, and prioritizing other city government programs over the fire department's when allocating capital investments.

CHAPTER 9

Scenarios for Elected Officials

1. Citizen Complaint: Transport Delay

▸ NFPA 1021: 6.2.1, 6.4.6

You receive an email from a citizen alleging that one of your fire department's ambulances, Ambulance 7, experienced navigation issues en route to the hospital, which resulted in delayed patient treatment. The email is from the patient's family. It states that the ambulance driver did not know where he was going and that, as a result of the long travel time to the hospital, the patient suffered adverse effects including delayed surgical intervention and spread of infection. The city's legal counsel was copied on the email message.

You asked the fire chief about the situation and requested a report from the dispatch center regarding the Ambulance 7 incident times for that particular call. The fire chief brings you the dispatch report, and a quick review of the computer-aided dispatch times revealed there was a 10-minute longer transport time than normal for the destination.

1. **If you do nothing, what will happen?**
2. **What do you say or do?**
3. **To whom do you say it?**
4. **What are two other alternatives you could use?**
5. **What, if any, legal issues should you consider?**
6. **What is your time frame for addressing this issue?**
7. **What is your last-resort action?**
8. **If you reach your last-resort action, what effect will you be dealing with two weeks after taking action?**

2. Fire Department Critical of City Administration

▸ NFPA 1021: 6.2.1, 6.4.6, 7.2.2

A vocal minority of fire fighters have been complaining to the press and making public statements about the city administration's poor treatment of fire department personnel. The complaints started many months ago in a fire station, spread to more stations and shifts, and over the past three months—although the responsible group is limited to a minority of the entire membership—they have increased in intensity. The group accuses the city administration of not valuing or respecting the fire fighters' work, citing modest pay raises of less than 2 percent, not hiring additional fire fighters to increase staffing levels, and deteriorating fire stations, fire equipment, and apparatus. The complaints have become increasingly public, with opinion pieces in the local newspaper and social media campaigns on Facebook and other social media sites dedicated to exposing the fire department conditions and the alleged lack of city hall support. The increased publicity seems to be designed to turn public sentiment against the elected officials and put pressure on you and your municipal colleagues to increase fire fighter pay, make renovations, and replace old apparatus.

During a regularly scheduled meeting with each of the department heads and the city manager, the city manager states she has had enough of the constant complaints and accusations from the fire fighters and tells you and the fire chief it is time to put a foot down and show the fire fighters who they really work for. She suggests that it might be time to reconsider staffing alternatives within the fire department that would provide the same service with fewer fire fighters per unit or rethink the priorities of the city's public safety needs. You know that the city manager's comments are borne out of frustration about the fire fighters' continued complaints and accusations and that she wants to ensure that the public perceives the city's elected and appointed officials as a strong and effective administrative body. You are also keenly aware the fire chief serves under your sole authority.

In the past 24 months you and other city officials have championed fire department improvements, including slight salary increases, the replacement of one fire engine and one ladder truck out of the 25 apparatus fleet, and renovations to the 75-year-old firehouse. You know that the city is facing significantly reduced revenues with no significant improvements within sight.

1. **If you do nothing, what will happen?**
2. **What do you say or do?**
3. **To whom do you say it?**
4. **What are two other alternatives you could use?**
5. **What, if any, legal issues should you consider?**
6. **What is your time frame for addressing this issue?**
7. **What is your last-resort action?**
8. **If you reach your last-resort action, what effect will you be dealing with two weeks after taking action?**
9. **What organizational impacts might the last-resort action have on the fire department?**
10. **What organizational impacts might the last-resort action have on the city administration?**

CHAPTER 10

Discussion for Elected Officials

1. Citizen Complaint: Transport Delay

▸ NFPA 1021: 6.2.1, 6.4.6

You receive an email from a citizen alleging that one of your fire department's ambulances, Ambulance 7, experienced navigation issues en route to the hospital, which resulted in delayed patient treatment. The email is from the patient's family. It states that the ambulance driver did not know where he was going and that, as a result of the long travel time to the hospital, the patient suffered adverse effects including delayed surgical intervention and spread of infection. The city's legal counsel was copied on the email message.

You asked the fire chief about the situation and requested a report from the dispatch center regarding the Ambulance 7 incident times for that particular call. The fire chief brings you the dispatch report, and a quick review of the computer-aided dispatch times revealed there was a 10-minute longer transport time than normal for the destination.

1. If you do nothing, what will happen?

The city will risk appearing unresponsive or aloof for not responding to a serious complaint about a reported substandard performance of city services. Further, in any future political activity or campaign within the city or beyond, an opponent could easily leverage this issue as evidence of inadequate leadership for the provision of basic emergency services in a timely fashion. Ignoring the issue will enable political opponents to paint a picture of indifference or lack of necessary leadership and oversight of city operations. The chief elected official will want to act, although such action may come through other personnel in the city structure in order to provide some political distance from the issue.

2. What do you say or do?

As the chief elected official, you will most likely want to have the complaint investigated by someone from the city's executive leadership, such as the city manager. This investigation could be delegated to the department head of the risk management department or legal department since the citizen's family has threatened potential legal action, implied by their having copied the city's legal counsel on the message. This step will insulate the city and the fire department should inquiries about the situation begin from the public or the media.

The city administration will be able to accurately respond to those inquiries with a brief and direct answer that the city is aware of the allegations and that they are being investigated. Note, however, that you will have no knowledge of the detailed findings until you are briefed by the investigator at the conclusion of the investigation.

The investigator assigned will need to contact the fire chief as soon as possible to identify the specific details surrounding the situation. It would be best to know when the event occurred, who was involved, why any delay occurred, and how such a delay could potentially occur within the fire department. It would also help isolate and possibly defend the case if the fire department provided a summary report of the frequency of similar response delays throughout the city within the past year or two and which personnel were involved. This may help to determine whether the specific crew members assigned to Ambulance 7 that day might have been at fault, whether the issue is more of a systematic concern, or whether it was an isolated event and not indicative of a broader issue. The fire chief should be tasked with providing some indication about what corrective actions will be taken to prevent future occurrences. Depending upon the results of the fire department findings, the lead investigator may decide to make similar inquiries to the police department to assess whether there are other systematic delays for city emergency services.

You may need to set a specific deadline for the fire department to deliver the data to the lead investigator so that you may make a recommendation for the next steps. This must happen rapidly in order to prevent any public perceptions of lack of responsiveness by the city organization or the chief elected official. The city administration and the fire department may have different definitions of a rapid response. The risk manager may want initial answers within one or two business days, which is reasonable with a standard Monday through Friday, 8-hour workday schedule. However, due to the structure of many fire department work schedules, wherein personnel may have 48 to 72 hours between shifts, a two-day initial investigation would possibly require the Ambulance 7 crew members' involvement when they are off duty. It is preferable, and will raise less speculation, to instead conduct the investigation while the crew is on duty. Note, however, that this may create prolonged delays, which the citizen complainant could easily interpret as unresponsiveness.

3. To whom do you say it?

The directions and concerns will be conducted in telephone or face-to-face conversations during the early stages of the process. These personal communications allow some flexibility to begin assessing the situation, while keeping notes, procedural decisions, and any potential conflicting information from becoming material to be used against the city, the fire department, or the individuals involved through open records requests. Findings of facts should be documented, but a conversation or notes taken early in the investigation may have information that is inaccurate or unsubstantiated yet could be used in a potential legal battle to persuade a jury that conflicts between initial notes or discussions not aligned with a final report somehow indicate a cover-up. Results and subsequent actions to those results will be documented, but the procedural steps or discussions about what happened and whom should be involved in the investigation will most likely be handled in face-to-face meetings with the chief elected official, the investigator, and the involved department personnel.

The legal team of the agency or city will draft a letter of acknowledgement of the formal complaint to communicate a timely response. The chief elected official is the appropriate signatory for such a letter, because this provides a more personal touch than a letter from a municipal department. The letter would likely acknowledge receipt of the complaint and indicate that steps are being taken to fully investigate the situation and that follow-up communication will follow.

4. What are two other alternatives you could use?

One alternative would be to have the fire department investigate the incident itself and share the results of the investigation within a reasonable time frame. This alternative would simplify the investigation and would give the fire department the benefit of the doubt that the allegations are unfounded. Although it is important to extend trust to members of local government, it could become problematic should political adversaries, media investigatory teams, or legal opponents claim that you relegated the investigation to the very department responsible for the delay. This may still be the best alternative to pursue, but be aware of the political risks for both yourself and the municipal organization as a whole.

A second alternative could be a hybrid measure of the two previous methods. You could assign the initial investigation to the police department, not for criminal charges but to leverage the municipality's existing investigative resources. This alternative would insulate you during the initial stages of the investigation while simultaneously protecting the fire department from having to conduct an internal investigation of its own department. Keep in mind that the relationship between the police and fire departments must be healthy and respectful. If there are tensions between the two over pay disparities, working conditions, or leadership differences, this alternative will likely magnify these already strained relationships, which could compromise the investigation and reduce the effectiveness of any of your subsequent recommendations or findings. The possible advantage of this alternative is that the investigation could be considered an active police matter and have some exemption from freedom of information requests; however, for this reason, a legal opinion should be sought prior to selecting this alternative.

5. What, if any, legal issues should you consider?

All elements in this scenario have the potential for legal consequences. Communications, actions, inactions, discipline, withholding of discipline, and even the investigative process can become scrutinized by the legal system. Frequent communication with your legal counsel team will be required to analyze each step prior to any action instead of after the fact.

6. What is your time frame for addressing this issue?

It is important to address the concern quickly but not hastily; a thorough investigation will likely take days and possibly even a few weeks. You should follow up with the complainant within about three weeks unless extenuating circumstances require prolonged investigation. Should a prolonged investigation be necessary, it is prudent to communicate an update to the complainant within the first two or three weeks. This allows the complainant to know the initial acknowledgment letter was not simply a dismissal tactic.

7. What is your last-resort action?

The last-resort action might involve apologizing for the delayed transport time. This should be carefully considered by the legal counsel, especially concerning specific word choice. However, if the initial investigation reveals an issue of substandard performance that can be prevented in the future, the last-resort action would be to recognize the issue, reassure the public that steps are in place to prevent its reoccurrence, and prepare for any consequences.

This process will likely not be pleasant and may have long-lasting consequences individually and organizationally.

8. If you reach your last-resort action, what effect will you be dealing with two weeks after taking action?

Organizationally, a legal action will likely be pending against the city fire department, in which the complainant is seeking damages of some sort. This has financial impacts should a judgment be issued against the city, as well as increased criticism for the city and the fire department services. You, the city, the fire department, and possibly the fire chief will also likely have image restoration issues with which to contend. Admitting imperfections in expected services is a bold move and can come with community anger and scrutiny of leadership and leadership's prior decisions and actions. Even when many things are going well and progressing, to publically admit a fault is humbling and risky because the impacts—positive, negative, or indifferent—are not fully realized until after the fact.

The impacts to the fire department can be significant as well. From a financial perspective, if the city incurs the impact of any legal judgment against either the city or the fire department, even if the funds come from city-wide accounts or insurance policies, the fire department budget will likely experience some reductions. Such reductions are not punitive but, rather, a matter of taking reduced municipal funds and stretching them further. Essential operations would continue, but new programs, the expansion of existing programs, or additional capital investments could be deferred well into the future to recover from the unanticipated financial setback.

The organizational impacts go beyond dollars, too. Some chief elected officials may consider drastic action within the fire department, such as replacing the fire chief or a change in lower level leadership directly associated with the program management in EMS or street operations. However, this measure may create new allegations and turmoil should the fire department membership start to feel wrongly attacked, unsupported, or lose leadership they admire and respect. This may perpetuate and even exacerbate the effects of the last-resort action.

2. Fire Department Critical of City Administration

▸ NFPA 1021: 6.2.1, 6.4.6, 7.2.2

A vocal minority of fire fighters have been complaining to the press and making public statements about the city administration's poor treatment of fire department personnel. The complaints started many months ago in a fire station, spread to more stations and shifts, and over the past three months—although the responsible group is limited to a minority of the entire membership—they have increased in intensity. The group accuses the city administration of not valuing or respecting the fire fighters' work, citing modest pay raises of less than 2 percent, not hiring additional fire fighters to increase staffing levels, and deteriorating fire stations, fire equipment, and apparatus. The complaints have become increasingly public, with opinion pieces in the local newspaper and social media campaigns on Facebook and other social media sites dedicated to exposing the fire department conditions and the alleged lack of city hall support. The increased publicity seems to be designed to turn public sentiment against the elected officials and put pressure on you and your municipal colleagues to increase fire fighter pay, make renovations, and replace old apparatus.

During a regularly scheduled meeting with each of the department heads and the city manager, the city manager states she has had enough of the constant complaints and accusations from the fire fighters and tells you and the fire chief it is time to put a foot down and show the fire fighters who they really work for. She suggests that it might be time to reconsider staffing alternatives within the fire department that would provide the same service with fewer fire fighters per unit or rethink the priorities of the city's public safety needs. You know that the city manager's comments are borne out of frustration about the fire fighters' continued complaints and accusations and that she wants to ensure that the public perceives the city's elected and appointed officials as a strong and effective administrative body. You are also keenly aware the fire chief serves under your sole authority.

In the past 24 months you and other city officials have championed fire department improvements, including slight salary increases, the replacement of one fire engine and one ladder truck out of the 25 apparatus fleet, and renovations to the 75-year-old firehouse. You know that the city is facing significantly reduced revenues with no significant improvements within sight.

1. If you do nothing, what will happen?

Inaction will create discord among the city's leadership because the city manager is frustrated by the fire fighters' actions yet the fire chief is not likely to support measures intended to reduce staffing, compensation, or postpone fire department improvements. Doing nothing will also leave the city and the fire department open to an increased volume of public attacks by the vocal group of fire fighters, who assert that the city elected officials are aware of the situation and are doing nothing about it.

2. What do you say or do?

First you need to isolate the problem and reduce the impact on the remainder of the city's operations. Instruct the city manager and the fire chief to meet with you in private to discuss the issue rather than in the forum of the city-wide meeting. In the separate meeting, remind both officials of the importance of productive cooperation for the sake of the city's leadership team as well as the efficient operations of government services. Communicate to the fire chief the reality of revenue reductions and the city's responsibility to maintain a range of local government services, emergent and nonemergent, for the public. Given the city's recent financial investment in fire department personnel, equipment, and facilities, the fire chief should be instructed to acknowledge and publicize the value of the city's investment in emergency services, specifically the fire department. Debate regarding the quantity of investment does not negate the fact that small pay raises, some apparatus replacements, and some fire station renovations have already occurred. Further instruct the fire chief to communicate the fiscal reality of the city in terms the fire fighters can understand. This is not meant as a threat but, rather, meant to ensure all fire personnel understand that publicly attacking their governing body is unwise and unhealthy for the fire department, the city, and its citizens.

Advise the city manager to develop a counter-media campaign to boost the city administration's support of the fire department and all emergency response services in the community. Examples might include "safe sidewalk" or healthcare/wellness initiatives. If necessary, review any upcoming fire department initiatives, and encourage the message within the fire department membership by rallying around past achievements as well as any possible future projects. The fire chief and members that the city manager identifies within the department should be present as a show of solidarity and support.

Lastly, instruct the fire chief to rein in some of the complaining fire fighters. Identifying and addressing such public expressions of displeasure will be challenging; however, the fire chief must understand the position of the municipality and fulfill the request to put a stop to the distractions.

3. To whom do you say it?

These discussions should occur privately between you, the city manager, and the fire chief. This allows the responsible parties to discuss the matter, determine assignments and actions, and set a course for future communications. It is important to have a communication plan to coordinate who is responsible to answer questions and provide information, should the media or the public request a statement about the situation. Give the city manager and fire chief direct instructions regarding whom to communicate with and what messages to convey to accomplish a positive outcome.

4. What are two other alternatives you could use?

One alternative would be to allow the city manager to manage the city operations as she sees fit. Given the city's financial constraints, the city manager could both reduce spending and achieve her objective of sending a message to the fire department by placing future projects and acquisitions on hold. This would risk some potential political fallout given the fire department allegations citing lack of city support. The reductions in capital investments, even if rooted in the city's financial concerns and competing city-wide priorities, would risk drawing more attention to the allegations of lack of support, which you would have to attempt to redirect.

For this alternative to be most effective, the fire chief must be allowed to communicate that fire department improvement freezes are directly related to the fire fighters' public complaints about the city administration. In this scenario, the complaining fire fighters view small pay increases and apparatus replacements and station renovations as insufficient rather than a step in the right direction. As such, when tough decisions need to be made, the chronic or public complaints actually influence decision makers against rather than in favor of future supportive measures. The city's financial hardships would be the cause of the freezes, but the choice to look first at the fire department for possible spending cuts would be a direct result of the fire department's lack of solidarity and support for the city government. Fire fighters must see that the consequences of public complaining actually set the department and its objectives back rather than move them forward. Such communications would ideally be verbal discussions and briefings to allow vocal tones, facial expressions, and gestures to assist in the communication of the entire message without being misunderstood. With such an emotionally charged topic, it may become easy for an email or memo to be interpreted in ways it was not intended due to the sender's word choice or how the recipient interprets the tone.

A second alternative would be to direct the fire chief and the city manager to meet regularly to create and implement an action plan that they support. This would force the city's upper-level leadership team to work together for the common good. This may require some form of facilitation, depending upon the degree of frustration and tension at the outset. This alternative encourages teamwork and mutual respect for differing perspectives within the city's operations.

5. What, if any, legal issues should you consider?

If the fire department complaints and accusations become loud enough, it is possible that a citizen or group of citizens might attempt to pursue legal action if they believe they have suffered a loss of property or inadequate emergency care due to improper deployment of basic public services to the community. This may be difficult to prove or disprove, but as the fire fighters' discontent increases, it can open the door to potential claims against the fire department, the municipality, or you as the elected official.

6. What is your time frame for addressing this issue?

The follow-up meeting to privately discuss the matter should occur within hours or possibly days. The resulting directives you give the city manager and the fire chief will likely take two to four weeks to complete. Any subsequent action steps should take place over the following months. It may be necessary to expedite future steps should elections be held or other important city initiatives be up for discussion or legislative action. As chief elected official, you want to prevent a fire department issue from becoming a distraction during debates about other unrelated municipal issues.

7. What is your last-resort action?

An absolute last-resort action may be to consider changes in the city's leadership team. The fire chief, the city manager, or both may need to be reassigned or released from service should there be differences that cannot be overcome. The advantage of political appointments by chief elected officials is that if changes are needed to keep the city's operations from becoming polarized or retaliatory, the chief elected official has the executive powers to enact them. Although it is generally not recommended to make organizational changes as a result of complaints and accusations, the chief elected official needs a successful leadership team in place to reliably and cooperatively lead city services. Although some of the discontent may stem from deeper within the structure of city departments, the heads of those departments must be able to productively guide and lead their subordinates. Failure to exhibit such guidance and leadership can come to rest squarely with the respective department heads.

8. If you reach your last-resort action, what effect will you be dealing with two weeks after taking action?

The first and most pressing effect would be conducting a significant job search to fill a key leadership team position or positions. This is a complex task and involves significant effort and time investment in addition to all the other daily city issues. Supporters of the released personnel will likely become vocal, claiming that their former leader was victimized and that you and/or the city official are responsible for dismissing talent from the local government organization. Replacing the fire chief would result in increased outcry from fire personnel of another instance of the city's alleged lack of support for the fire department. The complexity of the issue may increase as some of those vocal opponents begin to lobby for the individuals they believe to be a suitable replacement.

9. What organizational impacts might the last-resort action have on the fire department?

A change in leadership is never easy. Turmoil, uncertainty, inaccuracies, rumors, and a prolonged selection process will likely place the fire department in a self-preservation mode rather than a progressive mode. If the city manager is replaced instead of the fire chief, fire personnel may view it as a successful ousting of an unsupportive official and may be encouraged to continue voicing their complaints. Even if changes do not occur within the city's upper-level leadership team, fire department operations may be impacted with reduced staffing or benefits, delayed apparatus purchases, station repairs, renovations, and more. If funds were available to address some of these fire department concerns, they may be addressed. However, in economically challenging times, although financial hardships will impact departments city wide, the first negative effects will likely be felt by the least supportive or most disruptive department.

10. What organizational impacts might the last-resort action have on the city administration?

Divisive relations between government departments or leadership are a problem. When the public learns of the discontent through the media or word of mouth, it may call into question the city's ability to lead and govern, thereby reducing the city's credibility as a whole. When the city is negatively impacted, the community is as well. The community needs public safety protections, recreational opportunities, governance, transportation, utilities, and leadership. The chief elected official is ultimately responsible for managing and leading these competing interests, solving problems, and ensuring the public receives essential services. The chief elected official must not allow one senior leader, department, or perspective to outweigh the needs of the community at large. You have a vested interest in effectively working out differences without leadership changes or fault-finding missions.

Epilogue

Throughout this book, the emphasis has been placed upon you receiving this information either by your own observations or your direct report bringing a problem to you. To turn a completely different light on these scenarios, review the scenarios from the perspective that the issues were brought to your attention by a fellow officer, your peer, about a problem with your crew or shift. Now you must consider that what you do, how you do it, and how quickly you do it, are all being closely watched by the other officers, other members of the department, and maybe even your supervisors. Now what? Do you find yourself dealing with the scenarios differently? Size them up again.

Officers must understand the importance of sizing up station issues accurately and effectively. It will serve the officer, his or her members, the department, and the community well.

Courtesy of Jeff Barlow.

If you find yourself dealing with the scenarios differently, then ask yourself why. Does the fact that the message came from outside your daily circle of influence really change the fact that you must address the situation? This is a trap that ensnares many people. The speed and amount of follow-up communication regarding the scenario may be different, but if you seek to change the mechanics of how you handle the scenario, you may be handling the situation incorrectly. The facts you are provided have not changed at all, so why would you change the methodology of handling the situation based solely upon the method by which you received the information? Think about each scenario with the new perspective and again focus on the facts as they are provided.

Write out your responses and compare them with responses you wrote previously. Remember, practice makes perfect.

If you think about this text's various scenarios and various perspectives of the fire department organization, the officer's actions all share some common elements. This is where you can learn from the common themes to improve your critical decision-making skills for station issues. This is the reason, the same basic questions were asked of you throughout the text: If you did nothing what would happen? What do you say or do? With whom do you address the issue? What are two other alternatives you could use? What is your last-resort action? If you reach your last-resort action, what effect will you be dealing with in two weeks? Some common lessons from these rudimentary questions include the following.

First, you must recognize and embrace your responsibility as an officer to engage in situations as soon as you can. Do not leave issues for someone else to address and, likewise, do not leave them unaddressed hoping they will go away or resolve themselves. You are an officer, and you have a responsibility to engage. Officers supervise.

Second, though you have a responsibility to engage, recognize the importance of keeping your chain of command informed of any issues on which you are working. Such communications may be very brief and vague without descriptors of specific station, crew, or issue, but it allows your supervisors to know you are aware of your responsibilities and you are engaged in fulfilling those responsibilities. This is especially true if dealing with a potentially volatile issue as it will reduce the shock should developments occur requiring your supervisor's involvement.

Third, when engaging in issues, follow some basic concepts in dealing with people. Remain neutral; ask questions to gather information before forming any conclusions. Be discreet; when gathering information for decision making, use privacy to protect the innocent. If you are unsure who is innocent, presume everyone is innocent until the information you gather indicates otherwise.

Fourth, once facts are gathered do not make too little or light of the situation. If anything, fire service officers tend to under-document. If something occurs and you become involved, document the events, the facts, the outcomes, and expectations. This could be a memo for your files, a formal document for the personnel files, or disciplinary documents. Do not allow apparatus bay conversations, supervisory clarification of expectations, results of complaints, fact-finding missions, and other officer supervisory responsibilities to go undocumented. Officers will spend significant amounts of time dealing with issues repeatedly when there is no documentation to clarify expectations and outcomes.

Fifth, resist feeding the rumor mill. When faced with an issue you are investigating or counseling someone about, you will be faced with the myriad of questions about what is or is not happening to someone or about a particular issue. If you are unable to discuss the matter, leave it as a statement, "I am unable to discuss the matter." Your answers to even the most obvious questions trigger deductive reasoning, causing people to draw their own conclusions, as incorrect as they may be. Furthermore, your answers will become the basis for those conclusions. Resist the temptation to comment on anything you are unable to report upon.

Lastly, you are human and your experiences with the organization and the people within it will at times cause you to jump to conclusions based upon your historical knowledge. You will have to guard against yourself. Remember the unsinkable Titanic? The ship's designer was confident the design and construction made the ship unsinkable. Fire service officers are thinkers. You will use many skills as a leader, manager, coach, counselor, disciplinarian, teacher, and much more. When you become so confident in your skills dealing with station issues and the people you supervise, be careful not to "sink your ship." Critically analyze every situation anew. You will be a better officer for it, and your personnel will respect you for your pursuit of objectivity in the fulfillment of your officer responsibilities.

Size it up.

References

Brophy, John R., American Academy of Orthopaedic Surgeons. *Leadership essentials for emergency medical services.* Sudbury, MA: Jones and Bartlett Publishing, 2010.

City of Berkeley. (2008). *Fire Lieutenant Job Description.* Retrieved November 21, 2011, from http://agency.governmentjobs.com/berkeley/default.cfm?action=viewclassspec&ClassSpecID=105014

City of Denver. (2004). *Fire Lieutenant Job Description.* Retrieved November 21, 2011, from http://www.denvergov.org/Portals/85/documents/Job%20Description.pdf

City of Norfolk. (2001). *Fire Lieutenant Job Description.* Retrieved November 21, 2011, from http://www.norfolk.gov/human_resources/class_Specs/Fire%20Lieutenant.pdf

National Fire Protection Association. *NFPA 1021 Standard for fire officer professional qualifications.* Quincy, MA: National Fire Protection Association, 2009.

National Fire Protection Association, International Association of Fire Chiefs. *Chief officer principles and practice.* Burlington, MA: Jones & Bartlett Learning, 2012.